Self-Localization, Mapping and Coverage with Resource-Limited Mobile Robots

Dissertation

der Mathematisch-Naturwissenschaftlichen Fakultät

der Eberhard Karls Universität Tübingen

zur Erlangung des Grades eines

Doktors der Naturwissenschaften

(Dr. rer. nat.)

vorgelegt von

Dipl.-Ing. (FH) Marius Hofmeister

aus Kaiserslautern

Tübingen

2011

Tag der mündlichen Qualifikation: 29.04.2011
Dekan: Prof. Dr. Wolfgang Rosenstiel
1. Berichterstatter: Prof. Dr. Andreas Zell
2. Berichterstatter: Prof. Dr. Wolfgang Rosenstiel

Bibliografische Information der Deutschen Nationalbibliothek

Die Deutsche Nationalbibliothek verzeichnet diese Publikation in der
Deutschen Nationalbibliografie; detaillierte bibliografische Daten sind
im Internet über http://dnb.d-nb.de abrufbar.

ISBN 978-3-8325-2887-4

Logos Verlag Berlin GmbH
Comeniushof, Gubener Str. 47,
10243 Berlin
Tel.: +49 (0)30 42 85 10 90
Fax: +49 (0)30 42 85 10 92
INTERNET: http://www.logos-verlag.de

Meinen Eltern

Abstract

In this thesis, three fundamental issues of mobile robotics are addressed: self-localization, mapping, and area coverage. Self-localization denotes the robot's task to determine its position with respect to a given map. Mapping is the procedure to create such a map of the environment. In area coverage, one or multiple robots have to pass over the entire free space of a target area. The strategies presented in this work aim at deploying robots with limited resources. We define this lack of resources in terms of limited processing power, restricted sensor quality and accuracy, as well as inaccurate robot movements.

In the first part, we present and examine efficient techniques for visually localizing resource-limited mobile robots. The proposed algorithms are intensively tested on wheeled minirobots indoors, as well as on a flying quadrocopter outdoors. Several global image features are examined under different image resolutions. Furthermore, a practical solution to facilitate localization for wheeled minirobots by a compass is suggested that enables autonomous and efficient self-localization. Summarizing, best localization results could be obtained by means of a gradient-based image feature and the incorporation of a particle filter as a probabilistic method. A minimalistic pixelwise image comparison and the use of images at tiny resolutions further led to robust localization results.

A common benefit of small mobile robots is that they can be purchased at low costs that allow the parallel deployment of multiple exemplars. Due to the mentioned limitations of the robots, we suggest for the remaining tasks of the thesis the use of a heterogeneous robot team. Multiple minirobots are supported by a more capable service robot with state-of-the-art computational power and high-quality sensors through detection and teleoperation within line-of-sight. In this way, the strengths of two different types of robots can be combined and specific weaknesses are compensated. In the second part of this thesis, we develop a path planning method that leads to efficient and robust area coverage by means of the heterogeneous robot team. In contrast to related work, the suggested technique allows the coordination of robots without accurate navigation and localization capabilities in this task. Thereby, the algorithm is one of few variants which incorporate different types of robots in coverage.

Finally, in the third part of this thesis, cooperative visual mapping is conducted by the heterogeneous robot team in simulation and real-world experiments. The results show that, despite the inaccurate navigation capabilities of the minirobots, robust self-localization is possible on the basis of the created mapping dataset.

Concluding, we tailored advanced techniques of robotics to restricted platforms in order to find a trade-off between maximum performance and minimal resources. Therefore, the suggested strategies contribute to the future use of resource-limited mobile robots in various applications.

Kurzfassung

In der vorliegenden Dissertation werden drei grundlegende Themenkomplexe der mobilen Robotik behandelt: Selbstlokalisation, Kartierung und Abdeckung von Gelände. Selbstlokalisation bezeichnet die Fähigkeit, die eigene räumliche Position in Bezug auf eine gegebene Karte zu schätzen. Kartierung ist die autonome Erstellung einer solchen Karte. Die Abdeckung von Gelände schließlich umfasst das gänzliche "Abfahren" des Freiraumes einer Umgebung. Die in dieser Arbeit vorgestellten Verfahren ermöglichen den Einsatz von Robotern mit begrenzten Ressourcen. Als Ressourcenknappheit definieren wir neben der eingeschränkten Rechenleistung der verwendeten Prozessoren eine begrenzte Sensorqualität und -genauigkeit sowie unpräzise Navigationsfähigkeiten.

Im ersten Themenkomplex der Arbeit werden effiziente Techniken der visuellen Selbstlokalisation für Ressourcen-limitierte Roboter vorgestellt und untersucht. Die Tauglichkeit der vorgestellten Algorithmen wird dabei sowohl für zweirädrige Kleinroboter in einer Innenraumumgebung als auch für einen Flugroboter in einer Außenumgebung getestet. Eine Reihe von effizienten globalen Bildmerkmalen wird in Kombination mit unterschiedlichen Bildauflösungen auf ihre Eignung hin untersucht. Darüber hinaus wird ein praktischer Ansatz zur Vereinfachung der Selbstlokalisation mittels eines Kompasses für zweirädrige Kleinroboter vorgestellt, der eine autonome und effiziente Lokalisierung sicherstellt. Zusammenfassend kann festgehalten werden, dass die besten Lokalisationsergebnisse mithilfe eines auf Gradienten basierenden Bildmerkmals unter Einbeziehung eines Partikelfilters als probabilistisches Verfahren erreicht wurden. Ein minimalistischer direkter Bildvergleich und die Verwendung von winzig aufgelösten Bildern führten ebenso zu soliden Lokalisationsergebnissen.

Ein bekannter Vorteil von Kleinrobotern sind die niedrigen Herstellungskosten, die einen parallelen Einsatz mehrerer Exemplare auf günstige Weise erlauben. Aufgrund der erwähnten signifikanten Leistungseinschränkungen der Roboter schlagen wir in den weiteren Teilen der Dissertation zur effizienten Abdeckung und Kartierung eines Geländes eine heterogene Roboterteamstruktur vor. Mehrere Kleinroboter werden dabei von einem Serviceroboter mit leistungsfähiger Rechenkapazität und hochpräziser Sensorik in der Ausführung ihrer Aufgaben unterstützt, indem sie im Sichtbereich detektiert und ferngesteuert werden. Auf diese Art und Weise werden die Stärken unterschiedlicher Robotertypen kombiniert und limitierte Ressourcen kompensiert. Im zweiten Themenkomplex der Arbeit wird eine Pfadplanungsmethode entwickelt, die zur effizienten Abdeckung des Geländes mithilfe dieses Roboterteams führt. Im Gegensatz zu verwandten Forschungsarbeiten erlaubt die vorgestellte Abdeckungstechnik damit den Einsatz von Robotern ohne akkurate Navigations- und Lokalisationfähigkeiten in dieser Aufgabenstellung. Damit ist der Algorithmus eine der wenigen Varianten, die unterschiedliche Robotertypen in der Abdeckung von Gelände einbinden.

Im dritten Themenkomplex der Arbeit wird schließlich, aufbauend auf der vorgestellten Abdeckungstechnik, die kooperative visuelle Kartierung einer Innenraumumgebung mittels des heterogenen Roboterteams in Simulationsexperimenten und realen Experimenten durchgeführt und untersucht. Die Ergebnisse belegen, dass trotz der ungenauen Navigationsfähigkeiten der Kleinroboter eine robuste Selbstlokalisation mittels des erstellten Kartierungsdatensatzes möglich ist.

Zusammengefasst werden kann, dass fortschrittliche Methoden der Robotik auf limitierte Robotersysteme zugeschnitten wurden, um einen Kompromiss zwischen maximaler Leistung und minimalen Ressourcen zu erzielen. Die entwickelten Techniken leisten damit einen Beitrag zur zukünftigen facettenreichen Verwendung Ressourcen-limitierter Roboter in unterschiedlichen Anwendungsfeldern.

Danksagung

Der Prozess, der zum Verfassen einer Dissertation führt, ist in der Regel ein intensiver und zeitaufwendiger. Am Ende dieses Prozesses möchte ich an dieser Stelle meinen Wegbegleiterinnen und Wegbegleitern für ihre Hilfe und Unterstützung danken.

Meinem Doktorvater Prof. Dr. Zell gebührt an dieser Stelle der erste Dank für das mir entgegengebrachte Vertrauen und die große Zuverlässigkeit, die während der gesamten Promotionszeit zu spüren war. Meinem Zweitgutachter Prof. Dr. Rosenstiel danke ich für die freundliche Übernahme der Zweitkorrektur. Das kollegiale Arbeitsumfeld des Lehrstuhls werde ich vermissen; ich danke daher Klaus Beyreuther für die immerwährende technische Unterstützung und allen Kolleginnen und Kollegen für die sehr gute Arbeitsatmosphäre. Mit Sara Erhard, Kiattisin Kanjanawanishkul, Marcel Kronfeld und Philipp Vorst entstanden gemeinsame Publikationen. Maria Liebsch und Andreas Kuhn haben als meine ehemaligen Studienarbeiterinnen und Studienarbeiter ebenfalls einen Anteil am Erfolg dieser Arbeit. Bei den Aufbauarbeiten der c't-Bots halfen maßgeblich Tobias Heide und Werner Dreher. Korrektur gelesen haben diese Dissertation Dr. Michael Bayer, Philippe Komma, Marcel Kronfeld, Sebastian Scherer, Philipp Vorst und Karl-Engelbert Wenzel. Allen genannten Personen danke ich herzlichst für Ihre Hilfe! Besonders erwähnen möchte ich an dieser Stelle Philipp Vorst für seine unerschöpfliche Unterstützung zu jedem erdenklichen Thema, sowie Karl-Engelbert Wenzel für die angenehme Zeit als Zimmernachbar.

Meine Promotionszeit wurde finanziell gefördert durch das Ministerium für Wissenschaft und Kunst Baden-Württemberg, sowie im Anschluß durch ein Graduiertenstipendium der Friedrich-Ebert-Stiftung. Die Zeit als Stipendiat in der Friedrich-Ebert-Stiftung habe ich dabei als große Bereicherung empfunden und bin dankbar für die zahlreichen tiefgründigen bis philosophischen Gespräche, die mir damit ermöglicht wurden. Herzlich danke ich Marianne Braun für die angenehme und vertrauensvolle Betreuung meiner Förderzeit. Des weiteren danke ich Dr. Simon Wiest und allen Beteiligten der Hochschule der Medien, die mich zu dieser Promotion ermutigt haben.

Zuletzt danke ich meiner Familie und meinen engsten Freundinnen und Freunden. Ich bin glücklich, immer auf Euch zählen zu können!

Contents

Chapter 1

Introduction

1.1 Motivation

The autonomy of mobile robots has increased significantly in recent years. While in the field of manufacturing industrial robots have played a major role for many years, nowadays even mobile robots are becoming part of everyday life. Flying drones are able to provide firefighters with a view of the environment. Domestic robots support the residents in specific tasks. In supermarkets, prototypes of service robots can be deployed to guide consumers to desired products. Moreover, in areas that humans cannot access safely, partially autonomous agents are able to assist, e.g., in the demining of affected regions or in the exploration of foreign planets.

However, the successful execution of these tasks frequently requires the use of robots that possess extensive processing power and highly accurate sensor systems. On the other hand, miniaturization is a continuing trend in engineering and particularly also in robotics that leads to the development of small mobile agents that can be purchased at considerably lower costs, but which inherently suffer from limited resources. While in some areas the deployment of large-size robots remains a necessity, e.g., in cases where communication with humans is required, small mobile robots more frequently begin to take over common robotic tasks. One advantage of small robots is that they can easily be deployed in larger numbers due to their low costs. In this thesis, we mainly focus on wheeled robots with dimensions of 10-20 cm which in the literature are referred to as *minirobots*. At a very elementary stage, nowadays also micro- and nanorobots are developed, whose components are close to the scale of 1 mm and 1 μm, respectively. This progress leads to visions in which miniature agents work and investigate in parallel, robustly and fast. However, in reality there is a large gap in technology and knowledge to be bridged before these visions become true. Miniaturization bears a number of challenges for researchers and developers that have to be faced. Specifically, small mobile robots are *resource-limited* in various ways: Due to their size, payload is restricted. This leads to smaller power sources and, as a consequence, to less available processing power. Frequently, only microprocessors are used. Sensors provide less accuracy due to their small sizes. Additionally, wheeled minirobots often suffer from imprecise movements due to noisy odometry and wheel slip that comes with their lower weight. These challenges necessarily lead to the development of new techniques focusing on robustness and efficiency.

Figure 1.1 presents the robot platforms which we utilize in this thesis. We particularly experiment with the wheeled minirobots that have a diameter of 12 cm. This size class of robots provides a good compromise between maximal performance and minimal resources and thus has been shown to be suitable for a variety of tasks [165]: In swarm robotics, swarms of relatively simple agents collectively accomplish tasks that are beyond the capabilities of a single agent [29, 59, 60]. In the simulation and adaption of biological behaviors as well as in evolutionary robotics, minirobots are successfully deployed [149, 170, 186]. Since mobile robots embed elements from diverse fields such as mechanics, electronics, signal processing, programming and energy management, they are also suitable for educational purposes [1, 2, 157]. In the two major robot soccer federations, competitions also include the use of minirobots[1]. Further important scenarios for small robots are the monitoring of environmental conditions as mobile sensor nodes [108], or the inspection of machinery [57]. Recently, we also investigated and developed solutions in the control theory domain with minirobots [121, 122, 123, 124].

Despite this large number of research areas, we found that there is still a need for efficient and robust strategies concerning the fundamental capabilities of resource-limited robots. One of the key abilities of mobile robots is to navigate robustly through an environment. Therefore, three essential issues are addressed in this thesis that are related to mobile robot navigation: self-localization, mapping and area coverage. *Self-localization* denotes the robot's task to determine its position with respect to a given map. *Mapping* is the procedure to create such a map of the environment. In *area coverage*, one or multiple robots have to pass over the entire free space of a target area at least once. The proposed self-localization and mapping techniques make use of a camera as a sensor.

To investigate the self-localization approaches of this thesis intensively, in addition to the abovementioned minirobots, we also experiment with a quadrocopter as shown in Figure 1.1. Because flying robots can only carry a certain amount of payload, computational capabilities are also restricted on these platforms.

Since multiple robots are usually able to accomplish more complex tasks than a single robot and since resource-limited mobile robots can be purchased at low costs, we perform area coverage and mapping with a group of wheeled minirobots. However, to let multiple robots collaborate in a task, an elaborated cooperation strategy is usually required to achieve efficiency, e.g., in the allocation of subtasks and the distribution of information among the robots. In contrast to other works, we therefore decided to make the robots cooperate in a heterogeneous robot team: The service robot shown in Figure 1.1 possesses state-of-the-art computational power

---

[1]RoboCup Small Size League [3], FIRA Amiresot [4]

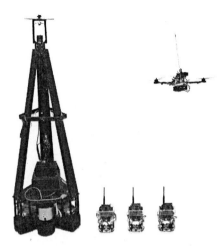

Figure 1.1: Robots deployed in the experiments of this thesis: a service robot with a height of approximately 1.5 m on the left, multiple minirobots and a flying quadrocopter.

and accurate sensors and is able to cooperate with the minirobots to fulfill their tasks. In this way, the strengths of two distinct robot platforms are exploited. Consequently, coverage and mapping algorithms are developed that make use of the heterogeneous robot team while inherently addressing efficiency in the fulfillment of the tasks.

Finally, the question arises why to use cameras for solving self-localization and mapping since a variety of sensors has been deployed in these tasks so far. One of the most well-known localization techniques is the Global Positioning System (GPS) that calculates a position based on measuring the time of flight of signals received from different satellites. However, in the standard version, the system provides an accuracy of only about 15 m. In addition, GPS works only in case of unobstructed line-of-sight to satellites and therefore not indoors. Thus, the most common sensor for localization in indoor environments on board robots has become the laser range finder. By measuring the time of flight of a circulating laser beam, the distance to objects can be determined accurately up to few millimeters. Though, laser range finders are usually expensive and too heavy to be mounted on small mobile robots. By contrast, camera systems have become inexpensive and lightweight devices. Particularly when using multiple robots, e.g., in swarm robotics or robot teams, sensor costs are an important factor to be considered. Furthermore, cameras are flexible sensors since they can help in a variety of tasks. Because of these reasons, we decided to use a camera as the main sensor in this thesis.

1.2 Outline

The following describes the outline of this thesis. We continue with a review of system architectures in Chapter 2. More specifically, we present a taxonomy of multi-robot systems and describe the architecture of our mobile robots. Additionally, the technical basis for the cooperation within our robot team is explained, that is, the detection and tracking of the robots in the team. In Chapter 3, we provide a comprehensive introduction and related work to self-localization techniques using a camera as a sensor. Our strategy to enable resource-limited mobile robots to localize themselves autonomously with focus on efficiency is subsequently described in Chapter 4. After that, we illustrate a technique to deploy a heterogeneous team of robots in area coverage in Chapter 5. Building on this, we establish cooperative visual mapping in an identical team of mobile robots in Chapter 6. Finally, the thesis is concluded with a summary and outlook in Chapter 7.

Chapter 2

System Architectures

To validate the applicability of our algorithms, both simulation and physical robot experiments are conducted in this thesis. Since in parts of this work we deploy multiple mobile robots in a team, we provide an overview and a classification of multi-robot systems in the first section of this chapter. Then, we technically describe our deployed robot platforms in Section 2.2. In our heterogeneous robot team, multiple wheeled and resource-limited minirobots work together with a custom-built service robot. To let these robots cooperate with each other, the service robot is enabled to detect and track the small robots visually by means of an omnidirectional vision system. The corresponding image processing techniques are described in Section 2.3, after outlining state-of-the-art tracking algorithms. Finally, Section 2.4 summarizes and concludes this chapter.

2.1 Multi-Robot Systems

Parker [183] describes the most common motivations for developing multi-robot system solutions: First, the task complexity might be too high for a single robot to accomplish it alone. Second, the task may be inherently distributed. Third, building several resource-bounded robots is much easier than having a single powerful robot. Fourth, multiple robots may solve problems faster using parallelism, and finally, the system's robustness increases through the redundancy of multiple robots.

The system architectures provide the infrastructure upon which multi-robot systems are implemented. They inherently determine the capabilities and limitations of the systems. Furthermore, they significantly impact on the systems' robustness and scalability. In the following, some key features of multi-robot architectures are discussed. Since this list can not be comprehensive, we refer the interested reader to the reviews [16, 43, 65, 183].

2.1.1 Centralization vs. Decentralization

Several philosophies of multi-robot architectures are imaginable. According to Parker [183], the most common types of architectures are centralized, hierarchical, decentralized and hybrid. Other authors categorize hierarchical approaches as a subgroup of decentralization and further define a subgroup of decentralization called distributed architectures [43].

Generally, in *centralized* architectures, the entire team is controlled from a single point. This structure is quite vulnerable since failures at a single point may lead to a collapse of the entire system. Furthermore, it can be unpractical due to the difficulty of communicating the entire system state back to the central location at frequencies that permit real-time control. However, the centralized architecture finds its relevance in situations in which the controller has a clear vantage point from which to observe the robots and from which messages to all of the robots can be broadcasted [128]. In *hierarchical* architectures, robots oversee the actions of a small or medium-sized group of other robots, each of which may oversee another group of robots, and so on, down to the lowest robot, which simply executes its part of the task. This architecture scales much better than centralized approaches. Its weakness is the vulnerability to failures high in the control tree. *Decentralized* (or *distributed*) approaches are the most common ones for multi-robot systems. The behavior of these systems is often described using terms as "emergence" and "self-organization". Typically, the robots take actions based only on their local knowledge. This strategy can be highly robust to failures, since no robot is responsible for the control of any other robot. Furthermore it is widely claimed that decentralized architectures have several advantages over centralized ones, including fault tolerance, reliability, exploitation of parallelism and scalability. However, Cao et al. [43] mentioned that there is no empirical or theoretical evidence of the claimed advantages. In addition, achieving high-level goals can be difficult since these goals must be incorporated into the local control of each robot. If the goals change, it might not be trivial to revise the behavior of all robots. Finally, in *hybrid* control architectures, local control is combined with higher-level control approaches to achieve both robustness and the ability to influence the entire team actions through global goals, plans, or control.

2.1.2 Homogeneity vs. Heterogeneity

A group of robots can be defined to be *homogeneous* if the capabilities of the individual robots are identical. This implies that the agents are composed of identical hardware and software. In contrast, *heterogeneous* robot systems are composed of varying behaviors, performances, size or cognition. This usually comes with a higher complexity since task allocation becomes more difficult and agents have a greater need to model other individuals in the group. However, complex applications may require the simultaneous use of multiple types of sensors and robots, all of which can not be included into a single type of robot. In several cases, heterogeneity may be a necessity to execute certain tasks, but it can also be a beneficial design feature: it may be more economical to enrich only a part of robots with specific capabilities rather than all [183]. Another reason to study heterogeneity is that it is nearly impossible to build exactly identical agents. Over time, even minor initial differences among robots, e.g., in construction and calibrations, lead to larger individual drifts.

2.1.3 Communication Structures

The communication structure in multi-robot systems determines the ways of interaction between the agents. Three major types of communication can be identified [183]. The simplest and most limited type of interaction occurs when the environment itself is the communication medium. In this case, there is no explicit communication or interaction between the agents. This type of communication is usually called *interaction via environment* or *stigmergy*. By contrast, in *interaction via sensing*, there is no explicit communication between the robots, but sensors are used to observe the actions of the teammates. Consequently, this technique requires that the agents are able to distinguish between other agents in the group or objects in the environment. Finally, there is *interaction via communication* in which robots directly and intentionally communicate relevant information by either directed messages or broadcast messages.

Dudek et al. [65] provided a more technical taxonomy of communication structures by specifying communication range, communication topology and communication bandwidth among other attributes. The *communication range* determines the maximal distance between two robots in a team such that communication is still possible. The *communication topology* describes the physical interconnections among the robots. Finally, the *communication bandwidth* indicates the amount of data which a communication link can transmit in a given period of time.

2.2 Robot Platforms

Various mobile robots are deployed in the experiments of this thesis. The investigation of visual self-localization is performed on the small, two-wheeled c't-Bots that are introduced in Section 2.2.1, and on an unmanned aerial vehicle, a Hummingbird quadrocopter, which is presented in Section 2.2.2. On these platforms, image processing is performed on separate modules: in case of the c't-Bot on a POB-Eye camera module and in case of the quadrocopter on a NOKIA N95 mobile phone. Both systems suffer from restricted computational power and limited sensor systems. To conduct visual mapping with multiple robots in this thesis, a

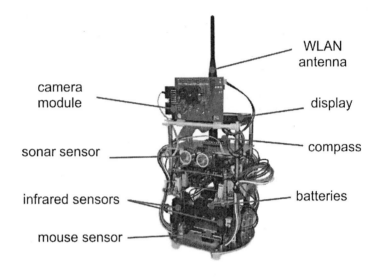

Figure 2.1: The c't-Bot platform in the extended version developed at our chair. The POB-Eye camera module was mounted on top of the robot to ensure the best possible viewpoint of the camera.

heterogeneous team of mobile robots is used. It consists of multiple c't-Bots and a custom-built service robot. After describing the service robot in Section 2.2.3, the structure of the robot team is introduced in Section 2.2.4.

2.2.1 c't-Bot and POB-Eye Camera Module

The differential-drive c't-Bot[1] was developed by the German computer magazine c't in 2006. It has a diameter of 12 cm and is 28 cm high, including the WLAN antenna (see Figure 2.1). This size class of robots is often referred to as *minirobots* [165]. Well-known robot platforms of similar sizes are, e.g., the Khepera robot [158], the s-bot [159], or the e-puck [157].

In the standard version, the sensor system consists of two Sharp GP2D12 infrared distance sensors, an ADNS-2610 optical mouse sensor for movement estimation, and several CNY70 infrared reflex light barriers for the detection of cliffs and ground lines. The wheel encoders provide a resolution of 60 counts per wheel. The deployed

---

[1]c't-Bot project, http://www.ct-bot.de

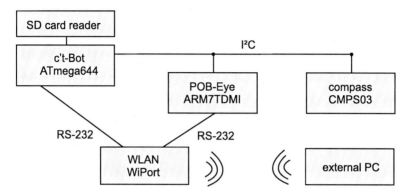

Figure 2.2: Overview of the c't-Bot architecture. While the SD card reader is directly connected to the microcontroller of the robot, the camera module and the compass communicate via I²C bus with the robot. The WLAN module is connected via RS-232 to the camera and the robot.

8-bit microprocessor is of type Atmel ATmega644, clocked at 24 MHz, with 4 KB SRAM and 64 KB flash program memory. To let the robot send and receive data, a WLAN extension board using a WiPort module was provided by the manufacturer that communicates by a serial port with the microprocessor. An SD/MMC card reader permits the storage of larger amount of data.

Additionally, the robots were equipped with several extensions at our chair. Because the maximum range of the Sharp sensors is only 80 cm, we mounted a Devantech SRF08 sonar sensor at the front of the robot that allows distance measurements up to approximately 2 m. A Devantech CMPS03 compass sensor with a declared accuracy of 3-4° was further added. Additionally, the c't-Bot community provides an open-source software framework in C that was used as a basis for our work. The underlying control architecture of the system is based on the subsumption architecture by Brooks [35], in which simple behaviors on multiple layers establish more complex capabilities for mobile robots.

Since image processing is usually computation-intensive, the need for a camera module arose that additionally provides computational capabilities. A well-known module is the CMUcam [196] developed by the Carnegie Mellon University. However, since version 2 of the CMUcam was not fully programmable, we decided on the POB-Eye camera module by POB-Technology[2] which is equipped with a 60 MHz ARM7 microcontroller programmable in C. We connected it via I²C bus to the ATmega644 and further provided the camera with a 2.1 mm wide-angle lens. The

---

[2]POB-Technology, http://www.pob-technology.com

GPS   IMU / microcontrollers

compass

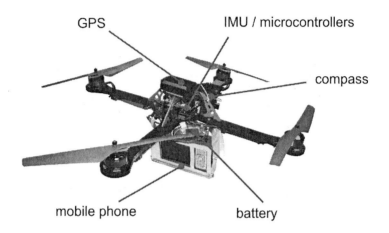

mobile phone   battery

Figure 2.3: The Hummingbird quadrocopter with Nokia N95 mobile phone that was put in a gondola. The inertial measurement unit (IMU) includes accelerometers and gyroscopes.

resulting opening angle was measured 96° horizontally and 64° vertically.

Figure 2.2 depicts the main components of the c't-Bot architecture. To visually self-localize as described in Chapter 4, image features are calculated on the POB-Eye module and are sent via I^2C bus to the robot's microprocessor. As the features of the previously grabbed training images can not be stored in the small RAM of the ATmega644, they have to be repeatedly loaded from the SD card for feature comparison. This makes the process both computation- and time-intensive.

2.2.2 Hummingbird Quadrocopter and Nokia N95 Mobile Phone

A quadrocopter is an aircraft that is lifted and propelled by four fixed rotors. By varying the speed of the four motors, the aircraft can tilt, turn and change its height. The X3D-BL Hummingbird AutoPilot quadrocopter distributed by Ascending Technologies (AscTec)[3] weighs approximately 0.5 kg and is 53 cm in diameter (see Figure 2.3). The platform includes two 32-bit ARM7 microcontrollers clocked at 60 MHz, a three-axis gyroscope, accelerometers, a compass module, a GPS sen-

---

[3]Ascending Technologies, http://www.asctec.de

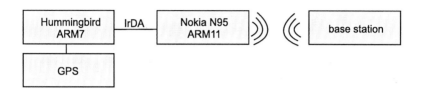

Figure 2.4: Overview of the quadrocopter architecture, where the mobile phone communicates via infrared with the quadrocopter.

sor and a pressure sensor. The PID control loop including inertial measurement unit (IMU) sensing runs at 1 kHz to stabilize the quadrocopter. The pressure sensor allows keeping a specific height, while compass and GPS can be used to keep a position in the air or to fly to waypoints. A ZigBee module enables a connection to a ground station. The detailed structure of the robot was described by Gurdan et al. [90].

Since the maximum payload of the quadrocopter is 200 g, a lightweight device is needed with the functionalities of a camera, computational power and memory. Furthermore, it must be able to communicate with the quadrocopter. The choice of Erhard et al. [75] was to use a Nokia N95[4] mobile phone. The overall system was named Flyphone. The N95 contains an ARM11 dual processor at 332 MHz, 160 MB internal memory and an 8 GB memory card. It further includes a 5-megapixel camera, a GPS sensor, and offers data transfer techniques such as WLAN, infrared and Bluetooth. The device is programmable in Symbian C++. Nevertheless, there are strong limitations in this embedded system referring to computational facilities (low throughput, no floating point support) and memory bandwidth (limited storage, slow memory, tiny caches) [75]. The images taken during flight missions are stored in an image database for later monitoring and the image features are stored in a relational database system. Since the mobile phone does not provide a serial connector, an infrared transceiver was attached to communicate via the Infrared Data Association (IrDA) protocol with the microprocessor of the quadrocopter. Figure 2.4 depicts the main components of the system.

2.2.3 Custom-Built Service Robot

The team of mobile robots which we experiment with in this thesis consists of multiple c't-Bots and a custom-built service robot. Formerly, the service robots had a height of approximately 0.9 m and were used in the Attempto Tübingen Robot

---

[4]Nokia, http://www.nokia.de/

Figure 2.5: Former structure of the service robots in the RoboCup environment, where at the chassis of the robots, a ball kicker was mounted [120].

Figure 2.6: Rebuilt structure of the service robots, where the ball kickers at the chassis were replaced by laser range finders [120].

Soccer Team playing in the Robot World Cup Initiative (RoboCup). They were developed and constructed by Heinemann [98] and the workshop of the Department of Astronomy of the Institute of Astronomy and Astrophysics at the University of Tübingen. Since soccer playing is a highly dynamic scenario, an omnidirectional drive system was developed: the robots are able to move in any direction without reorientation. This is achieved by means of three Swedish wheels which are mounted on the periphery of the chassis. In the old structure of the robots designated for the RoboCup competition, the sole sensor was an omnidirectional vision system. It consists of a Marlin F-046C camera by Allied Vision Technologies[5] pointing towards a hyperbolic mirror that is mounted on top. The Marlin F-046C provides

---

[5]Allied Vision Technologies, http://www.alliedvisiontec.com

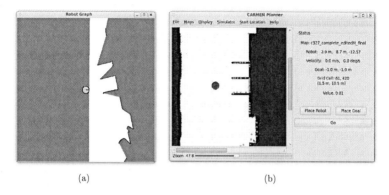

(a)   (b)

Figure 2.7: CARMEN graphic interfaces: (a) The robotgui module which illustrates
the current measurements of the laser range finder. It allows direct motion control of the robot. (b) The navigatorgui module that indicates
the current position and orientation of the robot with respect to a previously built map. It allows manually assigning the current robot pose
and the selection of waypoints to move to. The robot is depicted as a
circle, the short line indicates its front side.

an image resolution of 780×580 pixels and is able to capture and transmit 50 fps
at a reduced resolution of 580×580 pixels via the IEEE 1394a FireWire bus in the
YUV4:2:2 format [98].

Later on, the robots were rebuilt and now have a height of approximately 1.5 m to
potentially interact with humans. The main processing unit is a Pentium M 2 GHz
on-board PC with 1 GB RAM. The sensor system of the robots was extended by a
SICK S300[6] laser range finder and six Devantech SRF02 sonar sensors for collision
avoidance. The system uses two 24-Volt NiMH batteries. Figure 2.5 depicts the
old structure of the robots, Figure 2.6 the rebuilt architecture.

While in the former software architecture of the robot, a custom-made framework
was used [98], in this thesis, we work with the Carnegie Mellon Robot Navigation
Toolkit, named CARMEN [160]. CARMEN is an open-source collection of modular
software written in C and was designed to provide basic robot capabilities including:
base and sensor control, logging, obstacle avoidance, localization, path planning
and mapping. Most of the algorithms work with a laser range finder as a sensor.
Communication between the modules is handled using a separate package called
Inter-Process Communication (IPC) by Fedor and Simmons [5]. Figure 2.7 depicts
two graphical CARMEN modules.

---

[6]SICK, http://www.sick.de

Figure 2.8: Heterogeneous team of mobile robots, consisting of a service robot and three c't-Bots with colored hats.

2.2.4 Heterogeneous Team of Mobile Robots

As already stated in Chapter 1, we aim at exploiting the advantages of two different robot platforms by forming a heterogeneous robot team: Multiple resource-limited minirobots of type c't-Bot are deployed jointly with a custom-built service robot with state-of-the-art computational power and capable sensors. As the performance of these types of robots significantly varies, we propose a hierarchical team structure where the service robot oversees the actions of the minirobots and controls them.

The observation of the minirobots can be established by the omnidirectional vision system of the service robot. Theoretically, tracking could also be conducted by means of the laser range finder. The vision system, however, has the advantage to provide a 360°-view of the surrounding in contrast to only 270° in case of the laser range finder.

For the visual detection of the minirobots, a unique identification would be desirable, e.g., by individual patterns mounted on the minirobots [38] or by fiducials [185]. However, Kuhn [134] came to the conclusion that it is hard to uniquely identify our minirobots by means of the omnidirectional vision system. The provided image resolution at the enhanced height of the service robot may not be sufficient for this task. To let the robots be identified despite this restriction, a

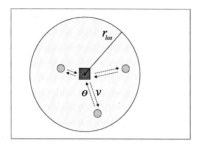

Figure 2.9: System structure of the heterogeneous team of robots. The three circles denote minirobots, the square illustrates the service robot. r_{los} is the line-of-sight radius of the service robot in which the minirobots can be visually detected and tracked. v represents the velocity values sent out by the service robot and θ denotes the compass measurements of the minirobots.

tracking algorithm was proposed. This implies that a unique identity has to be manually assigned to the robots when they are initially detected in the image. Then, the robots are tracked over time such that the correspondence between the robot's identity and the detected object in the image plane does not get lost. To facilitate and accelerate the detection, we equipped the robots with colorful orange hats as shown in Figure 2.8.

By applying visual tracking, some constraints arise: the minirobots have to stay in line-of-sight of the service robot to let them be tracked continuously. Additionally, a minimal distance between the small robots should be maintained in order to prevent possible tracking losses.

Besides their visual localization, the minirobots can be controlled via explicit communication. We establish TCP/IP-based communication channels in the Wireless LAN between the service robot and each of the minirobots. Since the service robot is not able to determine the orientation of the small robots visually, they continuously transmit their compass measurements. The individual steering commands for the minirobots are sent out by the service robot. Figure 2.9 illustrates the system structure.

2.3 Detection and Tracking of the Robots in the Team

This section focuses on the visual detection and tracking of the minirobots in our heterogeneous robot team. A related issue is the mapping of the pixel coordinates at which a robot has been detected in the image to the real-world coordinates of the robot. To do so, *geometry calibration* has to be performed that will also be addressed in the following. At first, an outline of object detection, tracking and geometry calibration techniques is given. In Section 2.3.2 we introduce the image processing algorithms which are applied in our robot team. Finally, the overall system is evaluated in terms of computation time and accuracy in Section 2.3.3.

2.3.1 Related Work

Object Detection and Tracking

Tracking can be defined as the problem of estimating the trajectory of moving objects in the image plane. In subsequent image frames, consistent labels have to be assigned to objects. Yilmaz et al. [255] provided a survey and classification of object tracking methods. They recognized strong challenges in the tracking of objects due to abrupt object motion, changing appearances of objects and scenes, non-rigid object structures and camera motion. In addition, objects and scenes were assumed to be partially occluded. Since a comprehensive survey would go beyond the scope of this thesis, this section only outlines the variety of approaches that can be suitable for object detection and tracking. Therefore, we follow the taxonomy of Yilmaz.

Generally, there is a strong relationship between the *representation of tracking objects* and the selection of the appropriate tracking algorithms. Objects can be represented by their shapes and appearances. Examples for object *shape* representations are points [211, 239], primitive geometric shapes [53], object silhouettes and contours [256] and skeletal models [12, 24]. The *appearance* of objects expresses itself in terms of, e.g., color or texture. Several approaches combine shape and appearance-based representation. For instance, a probability density of the object appearance can be computed from the region formed by the object's shape. These density estimates can be parametric, such as Gaussian [262] and a mixture of Gaussians [180], or nonparametric, such as histograms [53]. Other appearance-based representations are, e.g., templates that carry both spatial and appearance information [78] and active appearance models that are generated by simultaneously modeling the object shape and appearance [70].

Closely related to the representation of objects is the selection of appropriate visual *features* for tracking. Common features are color, edges, optical flow and texture. For example, texture features could be used to describe the appearance of objects. Edges, i.e., object boundaries, would be appropriate for contour-based object representations. The goal is to find features which are unique in the way that objects in the feature space can be distinguished robustly. Analogously, in Chapter 4 of this thesis, we extensively discuss the selection of efficient image features for visual self-localization.

Once the objects appear in the images, they have to be detected by an *object detection* algorithm to enable their continuous tracking. In case of uniformly-looking objects, this procedure is also frequently referred to as *blob detection*. Commonly, only information of a single frame is used by the object detection method. Another variant is to make use of the temporal information coming from a sequence of frames. In such a procedure, changing regions are highlighted by frame differencing. In the following, we briefly review some relevant techniques that are used to detect objects.

- *Point detectors* find interest points in images that reveal an expressive texture in their pixel neighborhood. *Interest points*, also referred to as *keypoints*, are widely used in different domains, e.g., also in visual self-localization. They often provide a certain invariance to the viewpoint of the observer and to illumination changes. Well-known interest point detectors are the corner detectors of Moravec [162] and Harris [92] and the scale-invariant feature transform detector (SIFT) of Lowe [146]. By comparing the feature *descriptors* extracted at the interest points, objects can be classified.

- *Image segmentation* techniques partition the image into perceptually similar regions. Every segmentation technique addresses two problems: the criteria to find good partitions and the method for achieving efficient partitioning [212]. Straightforward approaches classify the pixels by their color if the object appearance allows it [37, 98, 117]. Another approach called "mean shift clustering" forms clusters in the joint spatial and color space to detect objects [52]. Image segmentation can further be formulated as a graph partitioning problem, where the vertices of a graph, respective pixels, are partitioned into disjoint subgraphs [253]. After all, "active contours" evolve a closed contour to the objects' boundary such that they are fully enclosed [126, 262].

- In *background subtraction*, a representation of the scenes called background model is built. Significant changes in an image region of the background model in incoming frames are detected as moving objects. In other words, frame differencing of temporally adjacent frames is performed [193, 252].

- Finally, *supervised learning* mechanisms perform object detection by learning different object views automatically from a set of examples that were manually labeled before. Once the features have been chosen for this task, various learning approaches can be used, e.g., neural networks [197], adaptive boosting [240] or support vector machines [179].

Once an object could be successfully detected by one of the algorithms above, the *object tracker* generates the trajectory of it over time by locating its position in every image frame. The creation of correspondences between the objects across multiple frames is either included into the object detection step or can be done separately. In the first case, the object region and correspondence is jointly estimated by iteratively updating the object location and region information from previous frames. In the second case, possible object regions are obtained in every frame by means of the object detection algorithm, and then the object tracker corresponds objects across frames [255]. Yilmaz et al. defined three categories of object tracking methods: point tracking, kernel tracking and silhouette tracking.

- In *point tracking*, the detected objects are represented by points in the images. The association of the points is based on the previous object state which may include position and motion. An external mechanism to detect the objects in the frames is required. Correspondences can be achieved by using deterministic methods [239] or statistical ones as Kalman filters [34] or particle filters [21].

- In *kernel tracking*, objects are tracked by computing the motion of a kernel in consecutive frames, where the term "kernel" refers to the object shape and appearance. The motion is usually in the form of parametric transformation, such as translation or rotation [52, 78].

- *Silhouette tracking* estimates the object regions and uses the information encoded into the regions as appearance density or shape models. By matching the shapes or performing contour evolution, silhouettes can be tracked [111, 256].

Geometry Calibration

Once the objects have been detected in the image plane, their robot-centered real-world positions have to be calculated. To do so, a function that maps the coordinates $[x_p \ y_p]^T$ from the image coordinate system I_p to the two-dimensional coordinates on the ground floor $[x_r \ y_r]$ of the robot coordinate system I_r has to be determined.

In general, the determination of extrinsic and intrinsic camera parameters are referred to as camera calibration which is necessary to extract metric information from 2D images. Zhang [259] classified these techniques into two categories. In *photogrammetric calibration* techniques, the camera calibration is performed by observing a calibration object whose geometry in 3D-space is known with a good precision [77, 236]. These approaches require an expensive calibration and an elaborate setup. In *self-calibration* techniques, no calibration object is needed. By just moving a camera in a static scene and finding correspondences between multiple images, both the internal and external camera parameters can be recovered [93, 150]. While this approach is flexible, it might not always be reliable since there are many parameters to estimate [259]. Another camera calibration technique suggested by Zhang [259] requires the camera to observe a planar pattern shown at a few different orientations where the camera's motion needs not to be known. This method is included in several current implementations.

Since omnidirectional camera systems are composed of a camera pointing upwards onto a convex shaped mirror, camera calibration itself is not sufficient for such systems. By knowing the calibrated parameters of the camera, the exact position of the camera and the mirror, and the shape of the mirror, the mapping function could be derived geometrically. However, the relative position of the mirror to the camera may change during transport or maintenance. This can result in a completely wrong mapping. Scaramuzza and Siegwart [202] classified previous work on the calibration of omnidirectional camera systems into groups similar to the ones mentioned above. The first group includes methods which exploit prior knowledge about the scene, such as the presence of calibration patterns as in [44]. The second group covers approaches that do not require this knowledge, e.g., techniques which find point correspondences by including the epipolar constraints through minimizing an objective function [119, 153].

Another type of methods only roughly approximates the mapping function: According to Heinemann [98], most teams in the RoboCup competition manually collect the coordinates of objects in the robot coordinate system on the floor level and assign their corresponding coordinates in the image plane. By fitting a two-dimensional function through these correspondences, an approximate mapping can be achieved. Due to its practicability, we also follow this approach in this thesis, as described in detail in the next section.

2.3.2 Image Processing Algorithms

Detection and Tracking

In our work we build on the Open Source Computer Vision Library (OpenCV) maintained by Willow Garage[7] in version 1.1.0. The library provides a large number of algorithms for real-time computer vision. The blob tracking facilities of OpenCV, also called video surveillance facilities, were described by Chen et al. [47]. The intention of these modules is to track moving foreground objects against static or dynamic backgrounds. Conceptually, a video processing pipeline is established from

---

[7]Open Source Computer Vision Library (OpenCV), http://opencv.willowgarage.com

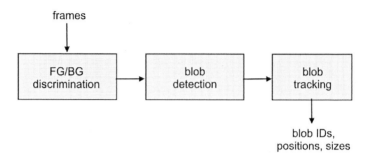

Figure 2.10: Blob detection and tracking pipeline in the OpenCV framework.

which we use the first three modules as depicted in Figure 2.10: a foreground/background (FG/BG) discriminator which labels each pixel as either foreground or background, a blob detector which groups adjacent foreground pixels into blobs, and a blob tracker which assigns identification numbers to blobs and tracks their motion from frame to frame. In the taxonomy of Section 2.3.1, the first two stages jointly form the object detection.

According to the requirements declared by the developers, the algorithm should work indoors as well as outdoors. Usually, the foreground detection is easier in indoor environments. In outdoor environments, dynamic backgrounds as wavering trees or flickering water surfaces make the discrimination more complex [47]. To tackle these challenges, two rather sophisticated FG/BG discriminators are provided in the framework. They base on the background subtraction techniques by Stauffer and Grimson [222] and Li et al. [141]. Stauffer and Grimson modeled each pixel as a mixture of Gaussians (MoG) and used an online approximation to update this model. The Gaussian distributions of the adaptive mixture model are then evaluated to determine which pixels are most likely to result from a background process. Li et al. formulated a Bayes decision rule for classification between background and foreground. Their method bases on pixel color and color co-occurrence statistics. For learning and updating the feature statistics of the background, learning strategies were proposed. Since this algorithm considers both moving and stationary background objects, we loosely refer to it as complex background handling (CBH) in the following.

As a drawback, the two provided background subtraction techniques make the first stage of the tracking pipeline both memory- and CPU-intensive. In our setup, the tracking frequency should be sufficiently high to also permit the tracking of moving robots from a simultaneously moving robot camera. We recall that our experiments will be performed indoors solely where we usually do not have to cope with dynamic backgrounds. Additionally, our minirobots can easily be equipped with colorful hats that facilitate their visual detection. For these reasons, we decided to additionally implement a straightforward color segmentation technique. It classifies each pixel by a pre-defined color look-up table as either foreground or background.

The second module of the pipeline, the blob detector, is inspired by a connected component (CC) tracker of Senior et al. [210]. It groups foreground regions into connected components. Then, the process tries to find each component in the previous frames by examining them on uniform motion and intersections. Components which are detected too near the image borders and components which exhibit an unreasonably high speed are not taken into consideration. If a blob could be tracked successfully across multiple frames, it is considered as correctly detected object and will be assigned a unique identification number by the blob tracking module.

Finally, the tracking module observes detected blobs over time while considering possible collisions between them. Originally, a hybrid tracker is proposed. If no object collision is likely to occur, a simple connected component tracker is used. It computes correspondences between objects in subsequent frames by their distances. A Kalman filter predicts possible collisions. If a collision is about to appear, a second tracker is used which combines the particle filter for tracking by Nummiaro et al. [171] with the mean shift technique of Comaniciu et al. [54]. However, since we will incorporate the collision avoidance among our robots in our motion planning algorithms and since the robots move in a single plane only, we do not strongly require the tracker to provide occlusion handling of objects. Therefore, instead of employing the hybrid tracker, we directly use the simple component tracker due to its low computational costs.

Figure 2.11 illustrates the detection of the minirobots in our omnidirectional vision system.

Geometry Calibration

In the following, we describe the geometry calibration, that is, the task of identifying the inverse perspective mapping function. Since the omnidirectional vision system of our robot was initially set up by Heinemann, the presented approach partially traces back to his work [98].

The inverse perspective mapping is a function

$$f_g^{-1} : \mathbb{N}^2 \mapsto \mathbb{R}^2, \begin{bmatrix} x_p \\ y_p \end{bmatrix} \mapsto \begin{bmatrix} x_r \\ y_r \end{bmatrix} \qquad (2.1)$$

that inverts the perspective mapping f_g by mapping the two-dimensional image coordinates $[x_p \ y_p]^T$ in I_p, originated in the image center, to the two-dimensional robot-centered coordinates $[x_r \ y_r]^T$ in I_r. Yet, the perspective mapping is only bijective and thus invertible for a specific two-dimensional plane, in this case, the

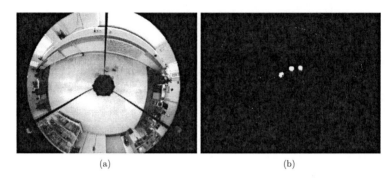

(a)   (b)

Figure 2.11: Robot detection and tracking: (a) omnidirectional image where the detected robots are marked by circles and labels. (b) color segmented image with blobs.

ground floor on which the robot moves. If points are detected that extend into the third dimension, they will be incorrectly mapped back as Figure 2.12 shows. Since we want to detect the colorful hats of the small robots that are in a certain height above the ground floor, we have to compute the perspective mapping error e to get the correct radius of the robots. This procedure is described below.

As the mirror of an omnidirectional camera system is usually rotationally symmetric, the inverse perspective mapping is also rotationally symmetric if the optical axis of the camera and the symmetry axis of the mirror are collinear and orthogonal to the floor plane, i.e., identical to the cross product $X_r \times Y_r$ of the two axes of I_r [98]. A precise production and mounting of our omnidirectional vision system resulted in a sufficient fulfillment of this assumption. By using polar coordinates and exploiting the symmetry and the angle preserving property of such a symmetric camera system, the two-dimensional inverse perspective mapping from Equation 2.1 can be simplified to

$$f_g^{-1} : \begin{bmatrix} r_r \\ \varphi_p \end{bmatrix} = \begin{bmatrix} f_r^{-1}(r_p) \\ \varphi_p \end{bmatrix},$$
(2.2)

where f_r^{-1} is the inverse perspective mapping of the radius in image coordinates in I_p to the radius in robot-centered coordinates in I_r:

$$f_r^{-1} : \mathbb{N} \mapsto \mathbb{R}, r_p \mapsto r_r$$
(2.3)

In this way, the task of calibrating the inverse perspective mapping is simplified to the identification of f_r^{-1}.

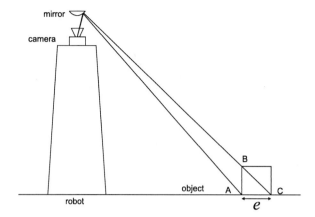

Figure 2.12: Illustration of the perspective mapping based on [98]. The mapping is only bijective for a two-dimensional plane, i.e., in our case the floor on which the robot moves. Objects that extend into the third dimension are not correctly mapped. The position of point A would be correctly computed, while point B would be incorrectly mapped onto the floor at point C.

Here, we conduct a manual calibration procedure that is similar to the semi-automatic method of Adorni et al. [11]. Several pairs of corresponding distances are collected, each pair consisting of the distance r_r of an artificial landmark to the robot in I_r and the distance r_p of its image to the image center. We approximate f_r^{-1} as $\widehat{f_r}^{-1}$ by fitting a polynomial of degree five through these pairs using a least-squares approach. The resulting polynomial is evaluated for each $r_p \in \{1, 2, ..., r_{max}\}$ and the resulting values are stored in a distance look-up table for efficient access. The maximum number of entries in the look-up table is 290 for the image resolution of 580×580 pixels on our robots.

The inverse perspective mapping can then be approximated by means of the center point of the mirror $[x_{p,c}\ y_{p,c}]^T$:

$$\widehat{f_g}^{-1} : \begin{bmatrix} r_r \\ \varphi_p \end{bmatrix} = \begin{bmatrix} \widehat{f_r}^{-1}(\widetilde{r}_p) \\ \widetilde{\varphi}_p \end{bmatrix} \tag{2.4}$$

with

$$\widetilde{r}_p = \sqrt{(x_{p,c} - x_p)^2 + (y_{p,c} - y_p)^2} \tag{2.5}$$

23

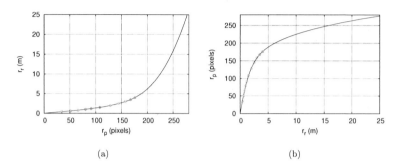

(a)   (b)

Figure 2.13: The manually calibrated mapping function $\widehat{f_r}^{-1}$ and its inversion $\widehat{f_r}$ for our omnidirectional vision system in (a) and (b), respectively.

and

$$\widetilde{\varphi}_p = \tan^{-1}\left(\frac{y_{p,c}-y_p}{x_{p,c}-x_p}\right) \tag{2.6}$$

Here, $\widetilde{r}_p$ and $\widetilde{\varphi}_p$ are polar coordinates in a frame parallel to I_p but centered in $[x_{p,c}\ y_{p,c}]^T$. This center point is also stored in the look-up table.

We now have to consider the tracking of our minirobots that have a certain height above the ground floor. As is formulated by Kuhn [134], the perspective mapping error e can be computed and compensated for objects with known size taking their smallest radius (cf. point B of Figure 2.12) as follows. Let h_o be the height of the detected object and h_r be the height of the robot with the tracking facilities. In our case, both values are known due to the fact that our "objects" are robots of equal and known size. Let furthermore $r_{\widetilde{r}}$ be the incorrectly determined radius (cf. point C of Figure 2.12). Then, by applying the intercept theorem, we can formulate

$$e = \frac{r_{\widetilde{r}} \cdot h_o}{h_r} \tag{2.7}$$

With the radius of the tracked robot r_{ct}, the radius r_r to the correct center point of the tracked robot can then be computed as

$$r_r = r_{\widetilde{r}} - e + r_{ct} \tag{2.8}$$

2.3.3 System Evaluation

To determine the performance of the presented tracking methods, we conducted tests on computation times and tracking accuracy. We compared the FG/BG discriminator based on the mixture of Gaussians (MoG) to the complex background handling (CBH) and to our straightforward color segmentation technique (CS) in combination with the proposed connected component tracker.

As expected, the incorporation of the more sophisticated FG/BG discriminators is computation-intensive: similarly as reported by Chen et al. [47], the use of the background subtraction techniques MoG and CBH lead to a computation time of 0.42±0.08 s and 0.34±0.02 s per cycle, respectively. This corresponds to a tracking frequency of 2.36 Hz and 2.91 Hz, which is rather slow for our scenario. By using our color segmentation technique, we achieve a cycle time of 0.04±0.01 s and thus a tracking frequency of 23.35 Hz. By scanning not the image as a whole but only parts of it, this frequency could further be increased. However, according to our experiments, the provided frequency was fast enough for tracking the small robots from a moving and slightly shaking camera. As mentioned above, we benefit from being able to cover the minirobots with unicolored orange hats that made our color segmentation algorithm feasible.

A well-known challenge for color segmentation techniques are changing illumination conditions that can make it hard to define a suitable look-up table. However, since our experiments are conducted indoors and we chose bright hats for the small robots, we discovered our look-up table to be robust enough to continuously track the robots, also under varying illumination conditions.

Another important property of the tracking system is the maximal distance at which robots can still be robustly detected in the omnidirectional images. Influencing parameters are the height of the camera system above the objects, the objects' size, the provided image resolution and the shape of the mirror. In our case, we discovered the tracking of our minirobots to work robustly within a line-of-sight radius $r_{los} = 2.0\,m$. To enlarge this radius, hardware modifications would be necessary, e.g., providing a larger image resolution, or mounting the camera closer to the ground. The averaged accuracy within $r_{los} = 2.0\,m$ was 0.04±0.02 m, measured at seven datapoints at varying distances from the robot.

Furthermore, in our experiments of Chapter 6 that lasted about 38 min, the tracking of the small robots was never lost. However, the false positive rate of detected robots was 7.63±6.84 during the entire experiment. This means that on average, eight times new robots were incorrectly detected in the images in addition to the correctly tracked ones. However, since these false detections were usually lost again soon, they had no negative influence on the correctly classified robots.

Finally, Figure 2.13 illustrates a manually calibrated mapping function and its inversion which we established for our service robot.

2.4 Summary

In this chapter, we introduced the system architectures of the robots we experiment with in this thesis. Two types of resource-limited mobile robots, a minirobot and an unmanned aerial vehicle, have been presented and technically described. They are deployed in the visual self-localization experiments of this thesis. We further provided a taxonomy of multi-robot systems since the coverage and mapping techniques of this work are developed with respect to a group of mobile robots. Specifically, multiple resource-limited minirobots are assisted by a more capable service robot to fulfill their tasks. Conclusively, our robot team was identified to be heterogeneous and hierarchical in the provided terms.

After all, we introduced the way the robots cooperate within the team. Besides explicit wireless communication, visual localization of the minirobots within the line-of-sight of the service robot was established. Different available tracking techniques have been tested with respect to their computation times. Since our experiments with the robot team are conducted indoors, and because we could equip the minirobots with colorful hats, we do not strongly require the use of sophisticated and time-intensive tracking algorithms. Instead, by applying a simple color segmentation, a higher tracking frequency of about 23 Hz was achieved. In the experiments of this thesis, this strategy was able to cope with a moving and eventually shaking camera while simultaneously tracking moving robots.

Chapter 3

Overview of Visual Self-Localization

The capability of determining a robot's pose with respect to a given map of an environment is generally referred to as localization. For mobile robots, self-localization is a key ability and a prerequisite to solve more complex tasks. This chapter aims at providing an overview of state-of-the-art self-localization techniques for mobile robots using the camera as a sensor.

In the first section, a general introduction and a terminology to this field of research is given. In most vision-based applications, the acquired image data is not directly used. Instead, different kinds of *features* are usually extracted from the images that serve for a more compact and distinctive representation of the essential image properties. Here, we distinguish between two types of image features: In Section 3.2, *local image features* are introduced for self-localization, followed by a description of *global image features*. Subsequently, a probabilistic technique referred to as Monte Carlo localization is derived in Section 3.3 that allows the estimation of the robot's pose in a more robust way by incorporating individual sensor characteristics and motion dynamics of the robot. Finally, Section 3.4 surveys related work on visual self-localization approaches.

3.1 Introduction

In general, three different kinds of localization problems can be distinguished [231]. *Position tracking* assumes that the initial pose[1] of the robot is known. By compensating the noise in the robot's motion, the actual pose can be estimated. In *global localization*, the initial pose of the robot is not known a priori, which makes the problem more difficult to solve. Finally, in the *kidnapped robot problem*, which can be regarded as a variant of the global localization problem, the robot gets kidnapped, i.e., the robot believes to know where it is from former sensor readings,

---

[1]For robots on the ground, *pose* denotes a two-dimensional coordinate in the plane together with an orientation value. In contrast, the term *position* neglects the orientation. In this chapter, we usually deal with poses rather than positions since this is the more general case, despite the fact that in the experiments of this thesis, we aim at estimating the position of the robot.

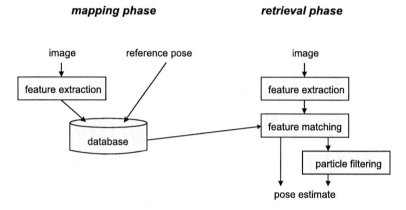

Figure 3.1: Illustration of mapping and retrieval phases in our visual self-localization approach, based on [246].

but it is in fact at another location. In this way, the ability to recover from global localization failures can be investigated.

Another way of distinguishing self-localization and also mapping approaches is the representation of the environment. In *range-based* representations, a two- or three-dimensional map denotes the outlines of the surrounding by modeling free and occupied space. In *appearance-based* techniques, the environment is represented by signatures of its appearance as color, shape or texture. Furthermore, the maps of the environments can be built in different ways: In *geometric* approaches, the robot's pose is determined metrically with respect to a Cartesian coordinate system. In *topological* models, an adjacency graph is created as a map representation, where nodes stand for distinct places. This concept is inspired from the way humans perceive the environment. The localization task on topological maps is referred to as place recognition. However, geometric maps usually provide a higher degree of accuracy.

In general, the visual self-localization techniques presented in this thesis are structured as *image retrieval tasks*. This means that our localization methods do not depend on specific environmental features such as, e.g., linear structures. Image retrieval is a more general process: The goal is to find the image in a database that reveals the highest similarity to a newly acquired image. This makes the approach flexible since various similarity measures or feature techniques can be applied. Usually, visual self-localization based on image retrieval comprises two phases, as depicted in Figure 3.1. In the *mapping phase*, images are acquired at points with known coordinates in an environment, which are referred to as reference poses. These coordinates can be determined for example by accurate GPS measurements where available, or manual measurements. Then, visual features are extracted from the images. These features are usually more compact than the image itself and describe essential properties of the image, e.g., textures or colors. The features are stored together with their reference pose in a database which forms a map of the environment. After a full map has been created, the robot is able to determine its pose with respect to this map. In this *retrieval phase*, also called *localization phase*, a new image is taken and features are again extracted. In the following, images which have been grabbed during the mapping phase are referred to as *training images*, while images acquired in the retrieval phase are referred to as *test images*.

By *matching* or, in other words, by comparing the features of a test image to the features of all training images, a simple pose estimate can be obtained: The reference pose of the features with the highest similarity can be directly taken. However, in case that two distinct places look similar, an estimate could be obtained which lies far away from the actual pose of the robot. To make the localization more robust to such kind of failures, probabilistical methods have been proposed. In this thesis, we use a particle filter for self-localization, also known as *Monte Carlo localization* [232]. In the terms of above, the self-localization strategies proposed in this thesis are appearance-based. Geometric maps are built to establish global localization. Examples of similar localization strategies are the ones of Wolf et al. [251] and Weiss [246].

One of the decisions that influences most the performance of the visual self-localization approach is the selection of suitable image features which are extracted from the images. Ideally, the features should be invariant or robust to the following changes which may occur in the images.

- *Translation and rotation:* If the test image is not grabbed at exactly one of the reference poses, this results in a translation in the image with respect to a training image. To detect the robot's pose regardless of this transformation, a feature has to cope with translations in the images. A rotation between test and training images usually occurs if the robot is not exactly positioned in the horizontal plane. In case of flying robots, this appears frequently. In case of wheeled robots, it happens on uneven ground, typically in outdoor environments.

- *Scaling:* If an object is observed from the same direction at different distances, this results in a scaled object representation. Although complete invariance to scaling is not desired, a certain robustness is useful.

- *Illumination changes:* The robustness to varying lighting conditions is a crucial aspect of a successful image feature. Illumination is likely to change due to varying light sources and weather conditions.

- *Occlusion:* In dynamic environments, moving objects result in the occlusion of certain parts of an image. Such partial differences of the images should be taken into consideration.

- *Noise:* Due to imperfect sensors, noise is likely to appear in the images, but this should not affect the self-localization abilities.

The literature generally distinguishes between two types of image features: *Global image features* describe the whole image as one single fixed-length vector. Well-known examples for such image features are color and grayscale histograms. By contrast, *local image features* describe only regions of high relevance in an image. One of the most famous local features is the scale-invariant feature transform (SIFT) developed by Lowe [146]. Due to their different structure, global and local image features have different advantages and disadvantages that will be explained in the following section.

3.2 Self-Localization as an Image Retrieval Task

Our self-localization approach can be considered as an image retrieval task. Consequently, this section describes how local and global image features can be successfully applied for finding the most similar image in a database to a newly acquired one.

3.2.1 Visual Self-Localization Using Local Image Features

Local image features are widely used in vision-based applications for object recognition [145], image retrieval [154] or building panorama images [36]. A desired property of local image features is their uniqueness, i.e., they should be highly distinctive. Besides the probably most popular scale-invariant feature transform (SIFT) of Lowe [146], a variety of local image features have been proposed. Examples for SIFT-variants are the speeded-up robust features (SURF) of Bay et al. [28] and the PCA-SIFT of Ke and Sukthankar [127]. Other local features detect corners in the image as the FAST corner detector by Rosten [194]. Also, performance evaluations of local descriptors and interest point detectors have been conducted [85, 155]. To understand the differences between local and global image features in more detail, we now introduce the basic concept of SIFT, since it is a standard in the community.

The overall aim of SIFT is to identify locations of *interest points*, also referred to as *keypoints*, that can be repeatedly found under different views of the same object or scene. After assigning an orientation to the extracted keypoints, a *descriptor* is calculated to define them by their pixel neighborhood. The resulting features offer invariance to image scaling and rotation and partial invariance to changes in illumination and the viewpoint of the observer.

(a)   (b)

Figure 3.2: Extraction of local image features of an image of size 320×240 pixels. In (a), the original image is shown. In (b), 593 extracted SIFT features are depicted. The origins of the arrows represent the location of the keypoints. The length of the arrows indicates the scales of the keypoints and the direction of the arrows represents the orientation of the keypoints.

Due to Lowe [146], four stages are required to generate a set of local SIFT features:

1. *Scale-space extrema detection:* After creating a scale space from the original image, potential interest points are identified that are invariant to scale and rotation. The scale space can be computed by increasingly blurring the original image.

2. *Keypoint localization:* The final keypoints are selected from the candidate locations computed above depending on their measured stability.

3. *Orientation assignment:* To achieve invariance to image rotation, an orientation is assigned to each of the keypoints based on the image gradient directions in their locality.

4. *Keypoint descriptor computation:* The final feature descriptor is constructed by incorporating the local image gradients at the detected keypoint scale.

Figure 3.2 shows an example image, in which SIFT features have been successfully extracted.

The best candidate match for a descriptor can be found by identifying its nearest neighbor in the feature space of all previously extracted descriptors. The similarity of a test and a training image may then be obtained by using the number of successfully assigned descriptors of the two images.

Figure 3.3: Extraction of a global image feature from an example image: A histogram of pixel intensities was built on the red color channel.

One of the main advantages of local image features is their robustness to occlusion. If objects are occluded partially, this does not affect the remaining extracted features. In addition, local image features have also been shown to be robust to changes regarding translation, rotation, scale and illumination. However, since many keypoints are usually found in an image, the computation of the numerous descriptors is time-consuming. Furthermore, in the matching step, the comparison of the features to all previously extracted features can be exhausting.

3.2.2 Visual Self-Localization Using Global Image Features

In contrast to local image features, which are extracted from areas of high relevance of an image, global image features describe the image as a whole. Well-known examples for global image features are color and grayscale histograms. A similarity value of two images can easily be obtained by directly comparing the two corresponding feature vectors. Figure 3.3 exemplarily illustrates the extraction of a global image feature.

In comparison to local image features, global features are more prone to modifications in the images since the whole vector of the image is then subject to change. However, one advantage of global image features is that their computation is commonly faster. In addition, also the matching step is significantly speeded up since only two vectors have to be compared. Finally, due to their compact representation, less storage space is needed to keep them.

3.3 Self-Localization Using Particle Filters

Since both sensor measurements and motions of a robot inherently suffer from uncertainties, probabilistic algorithms have been proposed to estimate the robot's pose more robustly over time. In position tracking problems, Kalman filters [118] have been successfully deployed [140, 205] due to the relatively small and incremental errors occurring in the problem over time. Kalman filters estimate posterior distributions of robot poses based on the collected sensor data. Uncertainties are represented by Gaussian distributions. However, the unimodal pose estimation makes it difficult to tackle global localization problems.

To overcome this limitation, multimodal approaches as multi-hypothesis Kalman filters or Markov localization have been suggested. Multi-hypothesis Kalman filters use mixtures of Gaussians to represent multiple distinct beliefs of the robot's position as in the work of Cox and Leonard [58]. Methods using Markov localization represent hypotheses of poses by piecewise constant functions over the space of all possible poses. An example for such a technique is the approach of Fox et al. [79]. However, one of the currently most popular probabilistic localization techniques is Monte Carlo localization (MCL) which was initially proposed by Thrun et al. [232]. By deploying a particle filter [64, 143] for robot self-localization, the global localization and kidnapped robot problem was shown to be solved in a robust and efficient way. In the computer vision community, particle filters are also known as condensation algorithms [112]. A collection of weighted particles is used to approximate arbitrary probability distributions. In contrast to previous methods, individual sensor characteristics, motion dynamics and noise distributions can be modeled. By modifying the number of samples, the particle filter can easily be adapted to robots with limited resources. Finally, particle filters are relatively easy to implement.

Since the MCL is a recursive Bayes filter which estimates the posterior of robot poses based on the collected sensor data, we subsequently describe the basic principles of Bayes filtering and illustrate then the approximation performed by the particle filter. The mathematical derivation of the Bayes filter follows the terminology of Thrun et al. [232].

3.3.1 Bayes Filtering

Bayes filters address the problem of estimating the state of a dynamical system from sensor measurements. In mobile robot self-localization, this state is the pose of the robot and the dynamical system is the robot in its environment. Measurements include range-sensor data, camera images or odometry readings. It is further assumed that the Markov assumption holds, i.e., that past and future data are conditionally independent if the current state is known. The posterior distribution, called the belief $Bel(x_t)$ is given by

$$Bel(x_t) = p(x_t|d_{0..t}), \tag{3.1}$$

where x_t is the state at time t and $d_{0..t}$ is the set of all previously acquired data. This data comprises robot observations $o_i, i \in \{0..t\}$, (i.e., measurements taken by cameras, laser scanners,...) as well as robot actions $a_i, i \in \{0..t-1\}$, (i.e., information concerning the robot's motion). Considering this, we can rewrite Equation 3.1 as

$$Bel(x_t) = p(x_t|o_t, a_{t-1}, o_{t-1}, a_{t-2}, ..., o_0) \tag{3.2}$$

It is assumed that observations and odometry readings arrive in sequence. Hence, a_{t-1} describes the robot's motion between time $t-1$ and t. By applying the Bayes rule, we can transform Equation 3.2 to

$$Bel(x_t) = \frac{p(o_t|x_t, a_{t-1}, ..., o_0) \, p(x_t|a_{t-1}, ..., o_0)}{p(o_t|a_{t-1}, ..., o_0)} \tag{3.3}$$

Since the denominator is constant regarding x_t, we replace it by the normalization constant $\eta = p(o_t|a_{t-1}, ..., o_0)^{-1}$:

$$Bel(x_t) = \eta \, p(o_t|x_t, a_{t-1}, ..., o_0) \, p(x_t|a_{t-1}, ..., o_0) \tag{3.4}$$

Applying the Markov assumption described above to the likelihood term $p(o_t|x_t, a_{t-1}, ..., o_0)$ yields

$$p(o_t|x_t, a_{t-1}, ..., o_0) = p(o_t|x_t), \tag{3.5}$$

which allows rewriting Equation 3.4 as

$$Bel(x_t) = \eta \, p(o_t|x_t) \, p(x_t|a_{t-1}, ..., o_0) \tag{3.6}$$

To obtain a recursive update equation, we now expand the right term by integrating over the state at time $t-1$:

$$Bel(x_t) = \eta \, p(o_t|x_t) \int p(x_t|x_{t-1}, a_{t-1}, ..., o_0) \, p(x_{t-1}|a_{t-1}, ..., o_0) \, dx_{t-1} \tag{3.7}$$

By applying again the Markov assumption, we can identify

$$p(x_t|x_{t-1}, a_{t-1}, ..., o_0) = p(x_t|x_{t-1}, a_{t-1}), \tag{3.8}$$

which then leads to

$$Bel(x_t) = \eta \, p(o_t|x_t) \int p(x_t|x_{t-1}, a_{t-1}) \, p(x_{t-1}|a_{t-1}, ..., o_0) \, dx_{t-1} \tag{3.9}$$

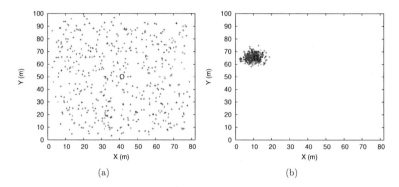

Figure 3.4: Position estimation using a particle filter. In (a), the initial state of the particle filter is denoted where the particles are randomly distributed over the target environment. In (b), the particle cloud has converged at a later point in time. The circle illustrates the weighted mean of the particles and thus the actual position estimate.

The final recursive update equation is then obtained by substituting Equation 3.1 into Equation 3.9:

$$Bel(x_t) = \eta \, p(o_t|x_t) \int p(x_t|x_{t-1}, a_{t-1}) \, Bel(x_{t-1}) \, dx_{t-1} \qquad (3.10)$$

Since Bayes filters estimate the belief recursively, there is a need for an initial belief of the state. In global localization, a uniform distribution over the state space is typically used.

For the practical implementation of the recursive update equation, two models have to be specified. The probability $p(x_t|x_{t-1}, a_{t-1})$ is usually referred to as *motion model* of the robot. It defines the probability of reaching a position x_t from a starting point x_{t-1} when the action a_{t-1} is executed. Frequently, odometry readings are denoted a. The second model, named *observation model* or *sensor model*, defines the probability $p(o_t|x_t)$, that is, the probability of making an observation o_t at position x_t. In other words, this step includes the prediction which observation would occur at state x_t and the comparison of this observation the actual one.

3.3.2 Approximation by Particles

In mobile robot localization, the state space is continuous. This makes the implementation of Equation 3.10 difficult, particularly in terms of efficiency. In the particle filter, the belief $Bel(x)$ is therefore represented by a set of m weighted samples

$$Bel(x_t) \approx \{x_t^{(i)}, w_t^{(i)}\}_{i=1,\ldots,m} \tag{3.11}$$

where $x_t^{(i)}$ is a *sample*, also called *particle*, of the random variable x at time t. Each particle can be recognized as a hypothesized position of the robot. $w_t^{(i)}$ are the weights, also called *importance factors*, of the particles at time t, i.e., non-negative values that sum up to 1. Considering a uniform initial distribution $Bel(x_0)$, the weights of all particles would be equally set to $\frac{1}{m}$. Figure 3.4 illustrates an initial and posterior particle distribution in position estimation.

Three steps are required to fulfill the recursive update in the particle filter:

1. *Resampling:* After the first iteration, m particles $x_{t-1}^{(i)}$ from $Bel(x_{t-1})$ are sampled according to their importance factors $w_{t-1}^{(i)}$.

2. *Prediction:* The motion model $p(x_t|x_{t-1}, a_{t-1})$ is applied to all of the particles to pass from state $x_{t-1}^{(i)}$ to state $x_t^{(i)}$.

3. *Correction:* The particles are newly weighted by incorporating the new observation o_t into the observation model $p(o_t|x_t)$.

Finally, after the weights have been corrected, they are normalized such that $\sum_{i=1}^{m} w_t^{(i)} = 1$. The current position estimate of the robot $\hat{x}_t$ can then be computed as

$$\hat{x}_t = \sum_{i=1}^{m} w_t^{(i)} x_t^{(i)} \tag{3.12}$$

The actual implementation of the particle filter in this thesis using global image features in the observation model is described in Section 4.4.2.

3.4 Related Work

This section gives an overview of research addressing self-localization techniques using a camera as a sensor. Researchers also focus on solving this problem jointly with mapping, which is referred to as simultaneous localization and mapping (SLAM). Such techniques are reviewed in Section 6.2 of this thesis.

Approaches to vision-based self-localization frequently differ in the type of deployed image features. In an early work of Horswill [106], different kinds of environmental-specific features related to walls, corridors or floors were extracted to realize navigation and place recognition. Basri and Rivlin [26] extracted edges and lines from images to build a geometric model for the training images. By fitting the data from newly acquired images via geometric transformations to the training images, a rough estimate of the scene could be determined. Another approach that converts image data directly into the corresponding camera pose without explicitly forming object models was presented by Dudek and Zhang [66]. They used a multi-layer neural network that was trained using backpropagation. Vertical edges have been used as visual features by Kortenkamp and Weymouth [132]. They further combined sonar and vision sensing to recognize places. Probabilistic self-localization based on maximum-likelihood estimation was suggested by Olson [176]. Occupancy grid maps generated by stereo vision were compared to previously generated maps to estimate the current position.

Sim and Dudek [216] described a technique to learn a set of image-domain models of selected features in a scene. Their method relates the appearance of a feature from a certain viewpoint to the pose of the observing camera and was successfully applied to global robot localization in a small indoor environment. Silpa-Anan and Hartley [215] used SIFT features at interest points detected with an adapted version of the Harris corner detector [92] to represent images. A kd-tree was used for indexing, matching and grouping a dataset to form a visual map database for localization purposes. Zhang and Kosecká [258] presented an image-based localization method in urban environments. After matching a test image to the training images by means of SIFT features, the closest view was found in the database. The fundamental matrix and homography was then used to estimate the motion between the two images.

Valgren and Lilienthal [238] compared SIFT, SURF and two of their variants in establishing outdoor localization on panorama images. Datasets of a period of nine months have been used, exhibiting large seasonal changes. Since the full invariance to rotation and scale of SIFT is not always desired, Andreasson et al. [14] developed MSIFT, the modified SIFT algorithm. By extracting these features from panoramic images, the robustness to localize in non-stationary environments was studied [15]. Another SIFT variant, called iterative SIFT, was proposed by Tamimi et al. [225, 226]. Their aim was to reduce the number of generated features as well as their extraction and matching time while still being able to accurately localize. Additionally, Tamimi et al. [225, 227] proposed the use of local integral invariants for localization, which have been originally presented as global image features [213, 251]. In the variant of Tamimi, the features were extracted at interest points found by SIFT. Recently, Ramisa et al. [190] studied combinations of different region detectors and descriptors to characterize places topologically. Their results showed that, if the number of detected features and possible matches increases, the localization accuracy could not always be improved. Different combinations exhibit different strengths and weaknesses depending on the situation.

In the context of RoboCup, localization approaches have been developed that make use of the typical properties of a soccer field. As an example, Heinemann et al. [98, 99] suggested to utilize line points of images and integrated them into Monte Carlo localization.

Several researchers used global image features for localizing mobile robots. Ulrich and Nourbakhsh [237] utilized color histograms extracted from omnidirectional images. By applying a simple voting scheme which incorporates each color band separately, place recognition is achieved. Zhou et al. [261] created multidimensional histograms to describe global appearance features of an image such as colors, edges or textures. In the retrieval phase, they limited the number of feature comparisons by considering spatial information within the topological map. Bradley et al. [33] focused on real-time outdoor localization using weighted gradient orientation histograms (WGOH). By dividing the image into a 4×4 grid of subimages, their feature vector became more robust to occlusion and was particularly evaluated under changing illumination conditions.

Another popular method to reduce the dimension of the image space is to apply principal component analysis (PCA) to the image data [19, 115, 116, 177]. Jogan et al. [116] proposed gradient filtering of the image eigenspaces to achieve extended robustness to illumination changes. Paletta et al. [177] used Bayesian reasoning to reject false hypotheses by incorporating spatial constraints. Tamimi et al. [228] first transformed the images into the wavelet domain and then applied PCA to the wavelet coefficients in order to reduce the feature dimension. Recently, Pretto et al. [189] described an image similarity measure based on the calculation of the Haar wavelet transform on gray-level images.

Wolf et al. [251] combined an image retrieval system, global integral invariant features and a particle filter to achieve geometric self-localization. An extension to global integral invariants was suggested by Weiss et al. [246, 247]. By dividing the images into a grid of subimages, the feature vectors become more distinctive and more robust to occlusion. The technique was called weighted grid integral invariants (WGII). Lamon et al. [135] developed fingerprint sequences that are extracted from panoramic images to uniquely identify locations. Specifically, fingerprints are a circular list of features that are encoded by a sequence of characters. Probabilistic self-localization for a mobile shopping assistant was examined by Gross et al. [88]. For each segment of a panoramic image, they extracted only the mean RGB-values as global image features. Prasser et al. [187] suggested a biologically-inspired feature extraction technique which makes use of Gabor filters. In a recent work of Murillo and Kosecká [166], the global gist descriptor [174] was adapted to match panorama images. Four gist words obtained from the test image were matched to the reference gist vocabulary. By taking the reference panorama that shares as many visual words as possible with the test image, place recognition was achieved. A comparison of SIFT features to global gradient histograms was conducted by Kosecká and Li [133]. They performed place recognition within a topological map and integrated spatial relationships by a hidden Markov model.

To obtain a similarity measure for the comparison of two existing feature vectors, several methods can be applied. Rubner et al. [198] provided a comprehensive overview of similarity measures for image retrieval tasks. They also proposed a new measure called earth mover's distance. Other investigations on comparing similarity measures for visual self-localization were conducted by Ulrich and Nour-bakhsh [237], Siggelkow [213], and by Erhard [74]. In the domain of localization by means of passive radio-frequency identification (RFID) fingerprints, Vorst and Zell [241] compared similarity measures.

In our localization strategy that will be presented in the next chapter, we reduced the image resolution to up to 11×15 pixels to speed up computation times. Small images for localization were also used in the early work of Argamon-Engelson [17]. By matching 64×48 images with different measurement functions, place recognition was performed. However, they did not compare the localization rates and computation times to other image resolutions. Our basic idea to reduce the image resolution was also inspired by the work of Torralba et al. [235]. They stored millions of 32×32 labeled images in a database and performed object and scene recognition on it. By using nearest neighbor methods as the sum of squared differences and variants, a reasonable performance on object recognition tasks was shown.

Finally, some researchers proposed ways to exploit the advantages of both global and local image features. According to Weiss [246], Artač and Leonardis [20] used global features to fastly select a small set of similar reference images to each test image. Then, they calculated SIFT features on these candidate images to obtain the final best match. Weiss et al. [246, 248] presented a hybrid localization approach that usually localizes by means of global image features. In difficult situations, e.g., under strong illumination changes, local features are used. To decide which type of features should be applied, the particle cloud of a particle filter used for position estimation is analyzed.

Chapter 4

Efficient Visual Self-Localization Techniques

After we provided a comprehensive introduction into visual self-localization techniques for mobile robots, this chapter now focuses on the challenging question how this task can be solved efficiently on resource-limited mobile robots. This chapter is an extended version of our related publications [101, 104, 105].

The self-localization strategies proposed here are extensively investigated on two different platforms in indoor as well as outdoor environments: on a small, two-wheeled minirobot and on an unmanned aerial vehicle, both initially presented in Chapter 2.

After presenting an overview of our self-localization approach in the first section, we address the way of dealing with the direction in which the images are grabbed in Section 4.2. Since choosing the right visual feature for self-localization is of great importance, suitable global image features are described in Section 4.3 that will be compared in the remainder of this chapter. Section 4.4 describes the way of estimating the robot's position. Two methods are proposed: After computing the image similarities, the best match can directly be taken as an estimate. A more robust way, however, is the incorporation of a particle filter. Its implementation based on the use of global image features is further explained in this section. Finally, the chapter is concluded with the illustration of our experimental results in Section 4.5 and a summary in Section 4.6.

4.1 Method Overview

As described in Section 3.1, our visual self-localization approach comprises two phases: a mapping phase and a retrieval phase. In the mapping phase, images are grabbed at reference poses in the environment. Since our small wheeled robot is not able to determine the ground truth by its own, we manually measure the indoor environment and put the robot at specified positions to grab training images. To achieve a complete coverage of the environment, we take the images at a grid of points. In case of the flying quadrocopter, we steer it manually on a trajectory and

Figure 4.1: Reducing the image size to lower resolutions. In the upper row, the resolution of an 88×120 image grabbed by the small wheeled robot is subsequently divided in half, up to a resolution of 11×15 pixels. In the lower row, a 320×240 image grabbed by the quadrocopter is downscaled up to a resolution of 20×15 pixels.

use the GPS measurements as ground truth for the reference poses. Despite the fact that ordinary GPS measurements may suffer from inaccuracies, we found them to be reliable in our scenario. However, since this assumption does not always hold, e.g., near buildings, the use of GPS as a sole sensor is not recommended. Therefore, vision is regarded as an important alternative for self-localization.

Since the decision of choosing the right feature for visual self-localization is crucial in terms of computation times and accuracy, we investigate the performances of different image features in this chapter. Because local image features require a large time to be extracted from the images and because the retrieval phase requires a lot of extensive feature comparisons, we chose to use global image features in the domain of resource-limited mobile robots.

Additionally, the time to extract features from an image heavily depends on the resolution of an image. We therefore investigate the impact of downscaling the images to tiny resolutions on localization accuracy and computation times. Figure 4.1 illustrates the reduction of the image size at example images of our localization experiments.

Nevertheless, comparing each feature of a newly acquired test image to all features of the training images can be time-intensive. Therefore, various methods to speed up image retrieval have been proposed so far [31, 71, 114]. However, we decided to incorporate a particle filter for self-localization as described in Section 3.3. It provides a good way to reduce the number of necessary feature comparisons. Moreover, arbitrary probability distributions of the robot's pose can be handled. In this way, the system can robustly recover from possible localization failures. The declaration of specific motion and sensor models furthermore addresses the uncertainties which occur in the motion and perception of the robot. Finally, the easy implementation and adaptation to existing resources make particle filters a good choice for resource-limited mobile robots.

4.2 Mapping Direction

Since in the case of our wheeled minirobots we grab images in a single direction only, in this section we discuss the way of dealing with the mapping direction. Generally, a popular sensor choice for visual localization and mapping is an omnidirectional camera [88, 200, 237, 238]. One reason that makes these systems attractive is the rotation invariance of the feature vectors that can easily be obtained. However, we equipped our robots with common cameras since omnidirectional systems also possess a number of disadvantages: first, such systems are often built at larger sizes, which makes them difficult to mount on small mobile robots. Second, omnidirectional cameras usually come at the cost of a lower resolution in the representation of objects. Third, the image representation of ordinary cameras corresponds to the perception of humans, which makes their use more intuitive.

Conclusively we have to consider the direction in which the test and training images are grabbed. Global image features do not facilitate this problem, since they are less robust to changes in the camera viewpoint than local image features. In a basic self-localization approach with common cameras, one would have to grab sufficiently many images towards several directions to map the area entirely. To circumnavigate this exhaustive process, researchers frequently restrict the directions of viewpoints, e.g., by moving the robots repeatedly on similar trajectories. This keeps the mapping direction approximately constant between the corresponding test and training images.

In case of our resource-limited wheeled minirobots, it is desirable to keep the number of training images small. Furthermore, manual mapping should not be too time-consuming. Since our aim is to make the deployed minirobots move autonomously in the environment, we here propose a practical way to specify the directions of viewpoints. By equipping the wheeled robots with an inexpensive and small compass module, it is possible to take training images only towards a single, defined direction. The robot is then able to rotate towards this direction to grab test images to self-localize. Due to magnetic deflections of furniture etc., the direction indicated by the compass is not always perfectly correct, but it is repeatable, apart from compass noise. Thus, the direction indicated by the compass can be seen as a function of the position. However, because of noisy odometry readings and wheel slip that come with the use of lightweight small robots, the robot's ability to control its movements and thus rotate towards a specified direction is restricted. Therefore, the actual robot orientations are likely to vary. This leads to approximate horizontal translations in the images that the image features have to

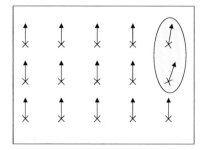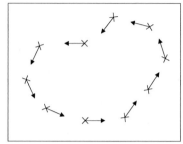

Figure 4.2: Different ways of dealing with mapping directions, depicted from top view. In the left example, training images are grabbed towards one constant direction on a grid of positions, where deflections may occur in the compass readings (as depicted in the ellipse). In the right example, the images are grabbed towards the direction of motion of the robot. Origins of the arrows depict the reference positions, arrows the mapping directions.

tolerate.

Summarizing, in case of the small wheeled robot, we grab images on a grid of points in the environment while looking towards one specified direction. In case of the quadrocopter, we make the common assumption that the robot flies repeatedly on similar trajectories and grabs images towards the direction of flight. Figure 4.2 illustrates the two kinds of dealing with mapping directions.

Wheeled Robots with Large Rotational Control Errors

In the abovementioned mapping technique which utilizes a compass, we expect that the wheeled robots come with a certain rotational control error. In fact, the rotational standard deviation of our minirobots was $\sigma_{rot_{ct}} = 8.79°$ in the mapping experiments of Chapter 6. The question arises how to tackle the case if $\sigma_{rot_{ct}}$ was significantly larger. In early experiments, we had to cope with $\sigma_{rot_{ct}} = 13.38°$ due to more imprecise motion control algorithms. Additionally, the horizontal opening angle of the robot's camera initially was not 96° as ultimately, but only 64° due to another type of lens. This complicates self-localization since, with a smaller camera opening angle, the image content changes a lot more under imprecise robot orientations. Therefore, we developed a strategy [104] that incorporates the robot's rotation impreciseness in the way the mapping is done. We grabbed multiple training images at each reference pose: one in the specified direction of mapping and four with a rotation difference of 10°, as depicted in Figure 4.3. In this way,

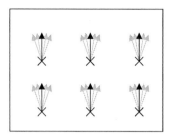

Figure 4.3: Extended mapping that addresses a large rotational control error of wheeled robots [104]. Origins of the arrows depict the reference positions, arrows the mapping directions.

we were able to establish self-localization despite the larger rotational control error of the small wheeled robots at a good accuracy.

However, because we were subsequently able to achieve a smaller rotational control error by implementing more accurate motion control algorithms and because of a larger opening angle of the camera lens, in this work we do not need to follow the extended mapping procedure, since it also requires more images to be grabbed.

4.3 Efficient Global Image Features

The selection of efficient global image features results from the limited computational power of our robotic platforms. In the following, we first present two more sophisticated algorithms for computing the feature vectors: *weighted gradient orientation histograms (WGOH)* and *weighted grid integral invariants (WGII)*. Both techniques showed good performance in earlier research and provided some robustness to illumination changes. Additionally, we also use traditional color and grayscale histograms for localization. We extended them by partitioning the images into a grid of subimages in the way it is done in WGOH and WGII. The idea behind this is to make the feature vectors more distinctive by adding spatial information. Furthermore, if parts of the image are occluded by objects, this only affects some of the subimages. Thus, only parts of the feature vector are distorted and not the vector as a whole. We tested the methods at different grid sizes and experimentally found out that a 4×4 grid is a good tradeoff between efficiency and retrieval accuracy.

To further decrease the feature extraction time, we downscale the images to tiny resolutions which hardly allow humans to recognize the image content anymore[1].

---

[1]According to Torralba et al. [235], humans need a minimal resolution of 32×32 pixels of images

More specifically, we preserve the aspect ratios of the images and interpolate the pixel intensities.

Motivated by the small amount of data that comes with the images at reduced resolutions, we also directly compare the images in a pixelwise fashion instead of first extracting a feature vector of the images. We call this technique *pixelwise image comparison* and describe it in the end of this section.

4.3.1 Weighted Gradient Orientation Histograms

Weighted gradient orientation histograms (WGOH) were first proposed by Bradley et al. [33] for localization in outdoor environments. They are inspired by the scale-invariant feature transform (SIFT) of Lowe [146] and are similar to the gradient orientation histograms proposed by Kosecká et al. [133]. In WGOH, each feature vector is created by firstly dividing up the image into an $m \times n$ grid of subimages. In this way, modifications that only affect parts of the image leave most of the feature vector unchanged. For each of the subimages, an 8-bin histogram is computed by weighted gradient orientations: At each pixel, the gradient orientations are weighted by the gradient magnitude and the distance from the center of the subimage. Gradients near the edges of each region are weighted less than gradients around the centers. The reason for this is that, if the image is translated slightly, gradients at the borders are more likely to fall into another subimage. Figure 4.4 illustrates the basic procedure of creating weighted gradient orientation histograms.

In case of $m = n = 4$, sixteen 8-bin gradient histograms are obtained which are concatenated into an 1×128 feature vector. After normalizing the vector to unit length, the size of any individual element of the vector is capped to 0.2, and the vector is normalized again. This step is analogous to the calculation of the SIFT descriptor and aims at reducing the dependence on few strong gradients or dominant regions of the vector.

In our version of WGOH that was adopted from Weiss [246] and that was prepared for the use on our robots by Liebsch [142], the practical implementation of the feature is as follows.

Let $A = \{A(i, j) \mid 0 \leq i < N_x, 0 \leq j < N_y\}$ be the original grayscale image of dimension $N_x \times N_y$ with $A(i, j) \in [0, 255]$. The gradients $G_x(i, j)$ and $G_y(i, j)$ in x- and y-direction are computed by convolving A with the Sobel operators S_x and S_y:

$$G_x(i, j) = A(i, j) * S_x, \tag{4.1}$$

$$G_y(i, j) = A(i, j) * S_y, \tag{4.2}$$

with

---

to reliably recognize objects.

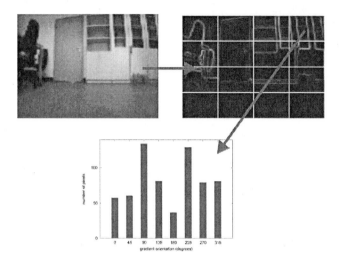

Figure 4.4: Extraction of weighted gradient orientation histograms based on [33, 246]. After a grid of subimages is created over the original image, for each of the subimages, a histogram of gradient orientations is calculated. To create the final feature vector, the histograms of subimages are concatenated.

$$S_x = \begin{bmatrix} 1 & 0 & -1 \\ 2 & 0 & -2 \\ 1 & 0 & -1 \end{bmatrix}, \; S_y = \begin{bmatrix} -1 & -2 & -1 \\ 0 & 0 & 0 \\ 1 & 2 & 1 \end{bmatrix} \tag{4.3}$$

The gradient orientation $\theta(i,j)$ and the gradient magnitude $m(i,j)$ at the pixel $A(i,j)$ are specified by

$$\theta(i,j) = atan2\left(G_y(i,j), \; G_x(i,j)\right) \tag{4.4}$$

and

$$m(i,j) = \sqrt{G_x(i,j)^2 + G_y(i,j)^2} \tag{4.5}$$

The gradient orientations at each pixel are weighted by their gradient magnitude and a 2D Gaussian by the distance to the center of the corresponding subimage, where the mean corresponds to the center of the subimage and the standard deviations correspond to 0.5 times the width and the height of the subimage, respectively.

4.3.2 Weighted Grid Integral Invariants

Global integral invariants have been initially proposed by Schulz-Mirbach [206] and have successfully been deployed in image retrieval [213, 214], texture classification [203, 204], and robot localization [246, 247, 251]. Essential improvements and enhancements to global integral invariants were suggested by Schael [203, 204], Siggelkow et al. [213, 214], and Weiss et al. [246, 247]. In our implementation and in this section, we adopt the weighted grid integral invariants (WGII) of Weiss et al. [246, 247], since this variant revealed good localization results even in outdoor environments under changing lighting conditions.

The idea behind integral invariants is to construct features which are invariant to Euclidean motion, that is, translation and rotation. Robustness to illumination changes is also a desired property of the features. To meet these requirements, all possible rotations r and translations (t_x, t_y) are applied to the image. The feature vectors are calculated by averaging over the transformed images. By using an appropriate kernel function as will be described below, integral invariants are also robust to local transformations as motion or transformation of individual objects.

Again, let $A = \{A(i,j) \mid 0 \leq i < N_x,\ 0 \leq j < N_y\}$ be the original grayscale image of dimension $N_x \times N_y$, where $A(i,j)$ represents the pixel intensity at coordinate (i,j). For an element g of a transformation group G, which is the group of Euclidean motions, the transformed image gA is given by

$$(gA)(i,j) = A(k,l),\qquad (4.6)$$

where

$$\begin{pmatrix} k \\ l \end{pmatrix} = \begin{pmatrix} \cos\phi & -\sin\phi \\ \sin\phi & \cos\phi \end{pmatrix} \begin{pmatrix} i \\ j \end{pmatrix} - \begin{pmatrix} t_x \\ t_y \end{pmatrix} \qquad (4.7)$$

Here, the rotation angle is given by ϕ and the translation in x- and y-direction by t_x and t_y, respectively. An invariant feature $F(A)$ for the image A can then be obtained by integrating over the transformation group G with a kernel function $f(A)$:

$$F(A) = \frac{1}{|G|} \int_G f(gA)\, dg \qquad (4.8)$$

In the discrete case of images, this can be rewritten as

$$F(A) = \frac{1}{|G|} \sum_G f(gA) \qquad (4.9)$$

By applying the group of Euclidean motions, the equation can then be extended to

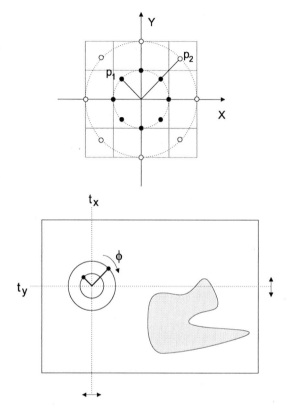

Figure 4.5: Calculation of global integral invariants based on [204, 213]. In the upper image, the feature value of the center pixel is computed by means of a relational kernel. The kernel incorporates pixels lying on different radii and phases around the center pixel. Two pixels p_1 and p_2 are shown with a phase shift of 90°. In the lower image, global integral invariants are computed for the entire image, where one feature value is calculated at each pixel. t_x and t_y illustrate the translation in x- and y-direction. ϕ denotes the rotation angle.

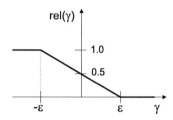

Figure 4.6: Definition of a ramp function for computing the relational kernel, where ε is a threshold value. The resulting values lie between 0 and 1.

$$F(A) = \frac{1}{RN_xN_y} \sum_{t_x=0}^{N_x-1} \sum_{t_y=0}^{N_y-1} \sum_{r=0}^{R-1} f(\, g(t_x, t_y, \phi = 2\pi\frac{r}{R})A\,), \qquad (4.10)$$

where R is the number of different rotation angles and ϕ is defined in the range $0 \leq \phi \leq 2\pi$. In this basic procedure, one single feature value $F(A)$ is computed for the entire image. A kernel function f is evaluated at regular intervals on a circle around each pixel. After summarizing the kernel values over all pixels (represented by the first two sums concerning t_x and t_y in Equation 4.10), the sum is divided by the total number of applied transformations. To make the resulting features more distinctive, instead of computing only a single feature value for an image, a histogram of feature values computed at each of the pixels is finally built.

Another extension to enrich the features is to incorporate multiple kernel functions and to form multi-dimensional histograms. Schael [203] proposed the use of relational kernels for being robust to illumination changes, because they calculate the differences between two pixels. The relational kernel for two pixels $p_1 = (i_1, j_1)$ and $p_2 = (i_2, j_2)$ is computed by

$$f(A) = rel\,(A(i_1, j_1) - A(i_2, j_2)) \qquad (4.11)$$

with the ramp function

$$rel(\gamma) = \begin{cases} 1 & \text{if } \gamma < -\varepsilon \\ \frac{\varepsilon-\gamma}{2\varepsilon} & \text{if } -\varepsilon \leq \gamma \leq \varepsilon \,, \\ 0 & \text{if } \gamma > \varepsilon \end{cases} \qquad (4.12)$$

where ε is a threshold value. Figure 4.6 graphically depicts this function.

Overall, the kernel functions compute the difference between the intensities of p_1 and p_2 lying on different radii and phases around the center pixel. The described procedure is repeated several times, where p_1 and p_2 are rotated around the center up to a full rotation while the phase shift is preserved. By averaging over the resulting differences, we get one value for each pixel and kernel, whereof the final histogram can be created. The basic procedure is illustrated in Figure 4.5.

The extension of Weiss et al. [246, 247] was to apply a subimage structure to the histogram similar to the one of WGOH (as introduced in Section 4.3.1): First, the image is divided into a 4×4 grid of subimages. For each of the subimages, an 8-bin histogram is computed, where pixels near the borders of the subimages are assigned lower weights, since they are more likely to fall into another subimage. A 2D Gaussian is used for weighting, whose mean lies at the center of the subimage and the standard deviations correspond to 0.25 times the width and the height of the subimage, respectively. After concatenating the histograms of the 16 subimages and normalizing them, we obtain a 1×128 feature vector by using one relational kernel, or a 1×256 feature vector by using two kernels, respectively. Note that by adding the subimage structure to integral invariants, the rotation invariance gets partially lost. However, we found the increased distinctiveness of the feature vectors to compensate this drawback.

In the final implementation, we set $\varepsilon = 0.098$, and we apply two relative kernel functions to each pixel. We experimentally found that the following radii for p_1 and p_2 lead to good results: radii 2 and 3 for kernel one and radii 5 and 10 for kernel two, each with a phase shift of 90°. One rotation is performed in ten 36° steps.

4.3.3 Color and Grayscale Grid Histograms

Color and grayscale histograms are well-known and efficient methods to solve a variety of tasks such as image retrieval [245], object detection [67, 224], or robot localization [88, 237] by representing the color distribution in an image or a subregion of it. In general, color and grayscale histograms can be built from any color space.

However, by forming a single histogram for the entire image, spatial information of the color distribution in the images is entirely lost. Images that look different can easily lead to similar histograms. Therefore we adopt the technique to form a grid of subimages as proposed in the previous histograms of gradients (WGOH) and integral invariants (WGII). We also utilize the specific weighting of pixels depending on their distance to the center of the subimage.

In practice, we use eight bins for each subimage in the histograms. Through concatenation we get a 1×128 feature vector of the 16 subimages. In case of the color histogram we use the hue value of the HSV color space. This choice of space promises to be quite robust to illumination changes.

4.3.4 Pixelwise Image Comparison

Global image features provide a compact representation of the essential image properties. Furthermore, they promise to be robust to changes in illumination, translation and rotation of the images. However, since we decided to significantly reduce the image resolution to save feature extraction time, a direct pixel comparison becomes feasible. This means that, instead of comparing the values of the histogram bins of two extracted feature vectors, we directly compare the pixel intensities of two images by means of a similarity function as introduced in the next section. In other words, the image data themselves are treated as a vector. To keep the amount of data small, we only compare grayscale images and discard color information. By previously normalizing the images, a certain robustness to illumination changes can be achieved. Additionally, since the opening angle of our wheeled robot's camera is enlarged by a wide-angle lens, some robustness to translations is inherently provided.

4.4 Position Estimation

In this section we first address the way of calculating the similarity between two feature vectors in order to obtain a position estimate. After describing how to derive a simple estimate by the *best match*, we illustrate the implementation of the particle filter operating on global image features to get a more reliable estimate.

4.4.1 Image Similarities and Best Match

In comparison to local image features, where multiple feature vectors have to be considered to obtain the similarity of two images, in case of global image features, the effort that has to be spent on the retrieval phase is low: Only two feature vectors have to be compared. According to our investigations, the normalized histogram intersection, which was originally proposed by Swain and Ballard [224], yields good localization results. In parallel, the computational effort is kept low as compared to other similarity measures such as cosine similarity or dissimilarity measures such as the Jeffrey divergence. Solely for the pixelwise image comparison, the normalized histogram intersection did not yield satisfactory results in our experiments. We therefore use the L_1-norm that can also be computed efficiently.

Accordingly, we calculate the similarity $sim(A, B)$ of two images A and B through normalized histogram intersection $\bigcap_{norm} (a, b)$ from the corresponding feature histograms a and b:

$$sim(A, B) = \bigcap_{norm} (a, b) = \frac{\sum_{k=0}^{m-1} \min(a_k, b_k)}{\sum_{k=0}^{m-1} a_k} \qquad (4.13)$$

Here, m is the number of histogram bins and a_k denotes bin k of histogram a.
The L_1-norm can be computed with the normalized images A^* and B^* as follows:

$$L_1(A^*, B^*) = \sum_{k=0}^{r-1} |A_k^* - B_k^*|, \tag{4.14}$$

where $r = N_x \cdot N_y$ is the number of pixels and A_k^* denotes pixel k of image A^*.
The similarity $sim(A, B)$ of two images can then be computed as:

$$sim(A, B) = 1 - \min(1, L_1(A^*, B^*)) \tag{4.15}$$

In general, $0 \leq L_1(A^*, B^*) \leq 2$, although $L_1(A^*, B^*) > 1$ rarely happens for images.

After the similarity of a test image to all training images has been computed, the training image with the highest similarity is then considered as the best match. Its coordinates can be directly taken as position estimate. More formally, let $\{(x_i, A_i)\}, i = 1, ..., n$, be n training images A_i grabbed at positions x_i. The position estimate $\hat{x}$ can then be obtained by means of the new observation, image B [241]:

$$\hat{x} = \underset{(x_i, A_i)}{\mathrm{argmax}} \; sim(A_i, B) \tag{4.16}$$

This procedure can be interpreted as a special case of *k-nearest neighbors (KNN)* with $k = 1$. However, since false matches are likely to appear at this stage, e.g., in self-similar environments, the particle filter presented in the next section incorporates the posterior probability and thus leads to more robust estimates.

4.4.2 Particle Filter for Self-Localization with Global Image Features

The particle filter has been theoretically derived in Section 3.3. This section now provides its practical implementation. In general, the belief $Bel(x_t)$ of the robot about its position x_t has to be approximated by a set of m particles. Each particle consists of a pose (x, y, θ) together with a nonnegative weight, its importance factor. The estimated position of the robot is given by the weighted mean of all particles. The initial belief is represented by particles which are assigned a random orientation θ and which are randomly distributed over the robot's global coordinate system. All importance factors are equally set to $\frac{1}{m}$. The particles are then updated for each test image iteratively in the stages *resampling, prediction* and *correction*, which will be described in detail in the following.

Resampling

After the first iteration, m random particles $x_{t-1}^{(i)}$ are drawn from $Bel(x_{t-1})$ according to the importance factors $w_{t-1}^{(i)}$ at time $t-1$. This step is only performed if the estimate

$$\widetilde{n}_{eff} = \frac{1}{(\sum_{i=1}^{n}(w_{t-1}^{(i)})^2)} \tag{4.17}$$

of the *effective sample size* [144] falls below the threshold $m/2$ that was empirically determined. The idea behind this is that the variance of the normalized particle weights is likely to increase over time. This, however, is not desired, since it implies that an increasing number of particles does not reasonably contribute to the probability density function close to the true position. Accordingly, resampling is performed.

Prediction

The sample $x_{t-1}^{(i)}$ is updated to sample $x_t^{(i)}$ according to an action a_{t-1}. In case of the small wheeled robot, we update the particles according to odometry and compass measurements. The odometry values measure the translation δ between two image recordings. The direction θ of the straight line motion is determined by the compass. Note that we perform only straight line motions with our robot between two image recordings. For each particle, Gaussian noise is added to δ with zero mean and standard deviation $\sigma_{trans} = \delta \cdot 0.1$. Additionally, Gaussian noise is added to θ with zero mean and $\sigma_{rot} = 21°$. We assigned a relatively large value to σ_{rot} due to the imprecise rotations of our robot. Furthermore, we set $\sigma_{rot} > \sigma_{rot_{ct}}$, since $\sigma_{rot_{ct}}$ (as measured in Section 4.2) does not incorporate the actual mapping direction with respect to the ground truth, but only the ability of the robot to rotate towards the direction indicated by the compass.

In case of the quadrocopter, no odometry readings could be obtained in our setup. Therefore, we let the particles disperse randomly according to a translation δ and orientation θ. δ is determined randomly by a Gaussian distribution with $1.0\,\mathrm{m}$ mean and standard deviation $\sigma_{trans} = 2.0\,\mathrm{m}$. To the orientation θ of each of the particles, Gaussian noise is added in each iteration with zero mean and $\sigma_{rot} = 45°$. These parameters correspond to typical velocities and flight behavior.

In the end, each particle is moved according to δ and θ.

Correction

The sample $x_t^{(i)}$ is weighted according to the observation model $p(o_t|x_t^{(i)})$. In case of the small wheeled robot, we first search the nearest training image $A(x_t^{(i)})$ to each particle. The current test image B and the training image $A(x_t^{(i)})$ are then matched by means of one of the abovementioned methods. The new weight is computed by

$$w_t^{(i)} = w_{t-1}^{(i)} \cdot sim(\ A(x_t^{(i)}), B\ ) \qquad (4.18)$$

This method is referred to as *standard weighting* in the following.

An alternative way of updating the particle weight is

$$w_t^{(i)} = w_{t-1}^{(i)} \cdot sim(\ A(x_t^{(i)}), B\ )^\lambda, \qquad (4.19)$$

with $\lambda > 1$. We refer to this method as *alternative weighting* in the following. By potentiating the original similarity, the differences between the weights become more distinctive [247]. This makes the particle cloud converge faster and focus on the particles with the largest weights.

The advantage of this observation model is that a maximal number of m feature comparisons have to be calculated. If the particle cloud has converged, an even smaller number of feature comparisons is required since many particles have identical nearest training images. This makes the model a good choice for resource-limited mobile robots. However, the underlying assumption is that images grabbed near each other look similar. Unfortunately, in the case of flying robots that suffer from unstable flight conditions due to air drafts etc., this assumption is not always correct. In this case, considering the best matches of all training images lead to more reliable and robust results.

Consequently, for our flying robot, we apply a sensor model based on *weighted k-nearest neighbors (WKNN)* similar to the one proposed for RFID-based localization by Vorst and Zell [241]. The special case of taking only the best match of all training images as position estimate (where $k = 1$) without using the particle filter was already described in Section 4.4.1. Here, we weight the particles individually by their distance to the weighted average of the k best matches to make the estimation robust.

Let again $\{(x_i, A_i)\}, i = 1, ..., n$ be n training images A_i grabbed at positions x_i and $sim(A_i, B)$ a similarity measure between a training image A_i and a test image B. Let further $\pi : \{1, ..., n\} \to \{1, ..., n\}$ be a permutation such that $\pi(i) \leq \pi(j) \Rightarrow sim(A_i, B) \geq sim(A_j, B)$. A direct weighted position estimate $\hat{x}$ can then be obtained by

$$\hat{x} = \frac{1}{\sum_{i=1}^{k} sim(A_{\pi(i)}, B)} \sum_{i=1}^{k} x_{\pi(i)}\ sim(A_{\pi(i)}, B) \qquad (4.20)$$

By incorporating this into the particle filter, we weight the particles individually by using their Euclidean distance to the weighted position estimate $\hat{x}$ by the following Gaussian:

$$w_t^{(i)} = w_{t-1}^{(i)} \cdot \exp\left(-\frac{1}{2}\left(\frac{||x_{t-1}^{(i)} - \hat{x}||}{\sigma_{wknn}}\right)^2\right), \tag{4.21}$$

where the standard deviation σ_{wknn} is the weighted distance between the k best matches and their weighted position estimate $\hat{x}$:

$$\sigma_{wknn} = \frac{1}{\sum_{i=1}^{k} sim(A_{\pi(i)}, B)} \sum_{i=1}^{k} sim(A_{\pi(i)}, B) \, ||x_{\pi(i)} - \hat{x}|| \tag{4.22}$$

σ_{wknn} gives also a statement of the reliability of the best matches. If σ_{wknn} is too large, we do not apply the sensor model (in our case, if $\sigma_{wknn} > 15.0\,m$).

The disadvantage of the WKNN pose estimate is that the test images have to be compared to all training images to obtain the k best matches. This obviously leads to larger computation times, but also to a higher robustness of the observation model.

Finally, after this correction step, we normalize the importance factors and calculate the estimated position by the weighted mean of all particles as in Equation 3.12.

4.5 Experimental Results

In this section we present the localization experiments and their results using the two robotic platforms in the in- and outdoor environments. After describing the experimental setups, we compare the performance of the proposed image features weighted gradient orientation histogram (WGOH), weighted grid integral invariants (WGII), color grid histogram (CGH), grayscale grid histogram (GGH), and pixelwise image comparison (PIC). Then, we describe the localization process with incorporation of the particle filter and present the corresponding localization results.

4.5.1 Experimental Setup

We conducted the experiments with the small wheeled robot in an office environment of approximately $75\,m^2$. Our training dataset consists of 190 training images that were grabbed facing a single direction with a manually oriented robot. The mapping direction was arbitrarily chosen west and was determined using the robot's on-board compass. The test dataset was grabbed at randomly chosen positions and consists of 200 images in total. 100 images, in the following called *test data A*, were grabbed at stable illumination. Another 100 images, *test data B*, were grabbed at different lighting conditions with and without the ceiling lights, at shining sun or dull daylight. In both datasets, the robot rotated autonomously towards west by

Figure 4.7: Example images of the datasets at changing lighting conditions. In the upper row, indoor images grabbed by the small wheeled robot are depicted. In the lower row, images of the quadrocopter are shown.

means of its compass. As mentioned above, this results in horizontal translations in the images due to the rotational control error of the minirobot.

To perform the experiments with the quadrocopter, we steered it at altitudes of around eight meters, flying rounds of approximately 180 m length in a courtyard. The view of the camera pointed in course direction. In total, we grabbed 1275 images in several rounds at a frequency of approximately one image per second. Each round consists of a different number of images due to the different velocity of the quadrocopter. In the mapping phase we grabbed 348 images at dull daylight. For the retrieval phase we used two different datasets: *Test data C* consists of 588 images (four rounds) at similar lighting conditions, *test data D* of 369 images (three rounds) at sunny daylight. Figure 4.7 shows example images of the different datasets.

4.5.2 Comparison of Image Features

Figure 4.8 reveals the mean absolute localization errors of the small wheeled robot using the different global image features at varying image resolutions. The localization error is the distance between the actual recording position of the test image and the corresponding best match, that is, the image with the highest similarity according to the feature comparison. In other words, the pose estimate of the robot is calculated by the KNN with $k = 1$, as formulated in Equation 4.16. Under the constant lighting conditions of test data A, the smallest mean absolute localization

Image data / feature	Size (bytes)
Image data (120×88)	10560
Image data (60×44)	2640
Image data (30×22)	660
Image data (15×11)	165
WGII	256
WGOH	128
CGH / GGH	128

Table 4.1: Storage requirements of image features and one-channel image data (grayscale) for the pixelwise image comparison. Note that the aspect ratio of the images is different for the wheeled and the flying robot. Exemplarily, we here illustrate the image resolutions of the wheeled minirobot.

error that could be achieved was 0.84 m using WGOH. The conditions of test data B led to a smallest mean absolute localization error of 0.93 m, again using WGOH. In Figure 4.9, the mean absolute localization errors of the quadrocopter are illustrated. These errors are measured in two dimensions only since we tried to keep the flying altitude constant. The smallest mean absolute localization error that could be achieved was 11.55 m on test data C and 13.36 m on test data D, both using WGOH.

Looking at the different methods in detail, we find out that in most cases WGOH led to best results (except at the smallest image resolutions). The results of WGII are worse, despite the fact that it also calculates differences of pixel intensities. Surprisingly, the pixelwise image comparison yielded constant results which are comparable to WGOH in case of the quadrocopter. In case of the small wheeled robot, the localization accuracy of the pixelwise image comparison is only degraded at larger image resolutions; at small resolutions, it even outperforms the more sophisticated methods WGOH and WGII. An explanation for this could be that the pixelwise image comparison does not average over the pixel intensities. While the grayscale grid histograms particularly under constant lighting conditions still provided reasonable results, a localization was not possible with the color grid histograms. Reasons for this may be the poor color quality of our cameras and the lack of meaningful color information in the environments.

As it could have been expected, the overall accuracy on test data B and test data D is worse, due to the challenging lighting conditions. Under those circumstances, the deployment of WGOH and the pixelwise image comparison is clearly advantageous in contrast to simple color and grayscale histograms. Furthermore, despite the significant reduction of the image resolutions, localization is still reasonably accurate.

Another important property of the features is their storage size. From Table 4.1,

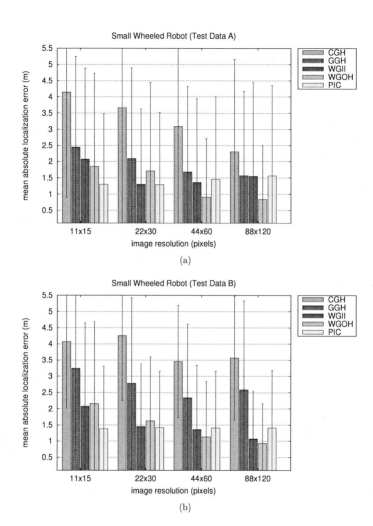

Figure 4.8: Mean absolute localization errors with standard deviations for the small wheeled robot using the best match. The different feature extraction algorithms are depicted at varying image resolutions on test data A and B in (a) and (b), respectively.

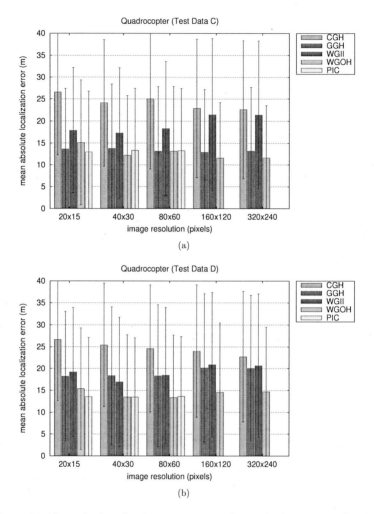

Figure 4.9: Mean absolute localization errors with standard deviations for the quadrocopter using the best match. The different feature extraction algorithms are depicted at varying image resolutions on test data C and D in (a) and (b), respectively. Due to database restrictions, no tests could be performed at the large image resolutions for the pixel-wise image comparison.

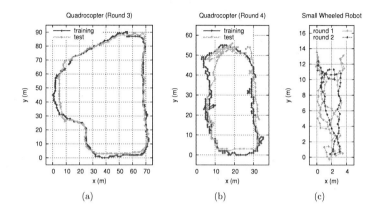

Figure 4.10: Real trajectories of the robots in the experiments using the particle filter. In (a) and (b) the test and training rounds of the quadrocopter are depicted. In (c) the two rounds of the small wheeled robot are shown.

it becomes apparent that the pixelwise image comparison which directly works on the image data is particularly feasible at small image resolutions for resource-limited mobile robots. The storage space needed for larger image resolutions is quite demanding.

In fact, the localization errors obtained so far are still quite large. However, the experiments mainly aimed at comparing the applicability of the features. By combining them with the particle filter in the next section, we expect a significant improvement concerning the localization accuracy.

4.5.3 Self-Localization Using the Particle Filter

To test the particle filter for localization on the small wheeled robot, we steered the robot arbitrarily two rounds through the environment as depicted in Figure 4.10 (c). To grab an image, the robot autonomously rotated facing west by means of the compass and afterwards rotated back to continue its path. Between the image recordings, the robot was steered straight ahead. *Round 1* consists of $n = 97$ images and was performed at stable lighting conditions. *Round 2* consists of $n = 62$ images that were taken at varying illumination conditions. The dataset of the previous section (190 images) was used as training dataset.

In case of the quadrocopter, we flew two different rounds, each of them two times. The training round of *round 3* consists of 238 images, the corresponding test round

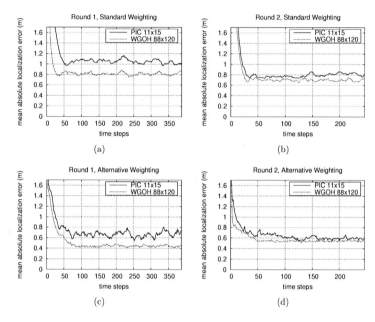

Figure 4.11: Mean absolute localization errors of the small wheeled robot using the particle filter with 40 particles. In (a) and (b) we applied the standard weighting to the particle filter. In (c) and (d) we used the alternative weighting (potentiating the similarity with $\lambda = 20$) for the different rounds.

of $n = 179$ images. In *round 4*, the training round consists of 348 images, the test round of $n = 142$ images. The real trajectories of the quadrocopter are depicted in Figure 4.10 (a),(b). In the WKNN-based sensor model of the quadrocopter, we set $k = 3$.

Then, we ran the particle filter on the test rounds, processing each round four times. To get the mean absolute localization error over time, we conducted this experiment n times; each cycle started at a different test image. Since our aim in using the particle filter was to keep the retrieval time short, we compared two different methods for this experiment on the small wheeled robot: the pixelwise image comparison on the 15×11 images and WGOH on the full-size images. We chose the pixelwise image comparison because it was the fastest method according to the feature extraction time while providing good accuracy. To compare the

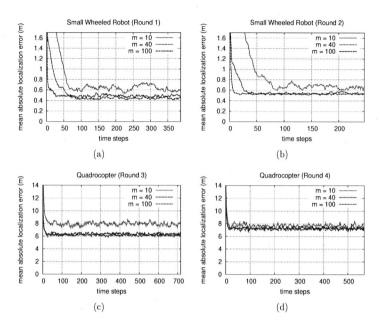

Figure 4.12: Mean absolute localization errors of the small wheeled robot and the quadrocopter on different rounds, where a varying number of particles was used for position estimation. WGOH was deployed on the full image resolutions and in case of the wheeled robot, the alternative weighting was used.

results, we used WGOH since it revealed the smallest localization error and a feature extraction time that was significantly shorter than WGII. On the quadrocopter, we deployed WGOH at the full resolution.

In Figure 4.11, we investigated the performance of the standard and alternative weighting methods introduced in Section 4.4.2 on the small wheeled robot. In the alternative weighting, potentiating the similarities with $\lambda = 20$ led to best results. Because the restricted memory of the robot allowed a maximal number of 40 particles to be calculated, we set $m = 40$ as the default value in these experiments. Due to the random initial distribution of the particles in the environment, the initial localization error is large: 3.77 m (3.52 m) in round 1 and 2, respectively. However, in all experiments, the particle cloud converged within the first of the four cycles of the rounds. As expected, the localization accuracy of the pixelwise image compar-

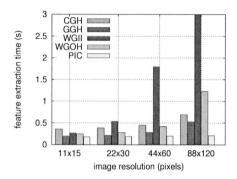

Figure 4.13: Feature extraction time for the small wheeled robot using different global image features under different image resolutions. The extraction time of WGII at 120×88 pixels is 7.03 s and had to be cut off for better visibility.

ison on the small size images was always worse than by using WGOH on the full resolution images. Nevertheless, a localization was still possible at a reasonable accuracy. In case of the standard weighting, the mean absolute localization error and standard deviation over all four cycles was 0.88±0.36 m (1.16±0.39 m) in round 1 and 0.82±0.45 m (0.91±0.44 m) in round 2, using WGOH and the pixelwise image comparison, respectively. In case of the alternative weighting, the corresponding error was 0.52±0.27 m (0.74±0.27 m) in round 1 and 0.60±0.23 m (0.67±0.24, m) in round 2, using WGOH and the pixelwise image comparison, respectively. This corresponds to a reduction of the localization error by using the alternative weighting of 40.9 % (26.8 %) in case of WGOH and of 36.2 % (26.4 %) in case of the pixelwise image comparison, in round 1 and 2.

Figure 4.12 (c),(d) depicts the localization results of the quadrocopter using the particle filter on two different rounds. Using $m = 40$ particles, the overall mean localization error was 6.43±1.24 m (7.24±0.75 m) in round 3 and 4, respectively. The corresponding initial localization error was 38.02 m (22.63 m).

To investigate the impact that the number of particles has on the localization accuracy, we conducted further experiments once with $m = 10$ and then with $m = 100$ particles (see Figure 4.12). The use of 100 particles did not always improve the localization accuracy. In contrast, deploying only 10 particles always lead to significantly worse results. The mean localization errors with $m = 10$ of the small wheeled robot are 0.81±0.49 m (0.96±0.66 m) and of the quadrocopter 8.05±1.38 m (7.73±0.76 m) in the two different rounds. In other words, by using only 10 instead of 40 particles, the mean localization error increases by 35.8 % (22.9 %) in case of the wheeled robot and by 20.1 % (6.3 %) in case of the quadrocopter in the two rounds. From these results, we believe that $m = 40$ is a good tradeoff between efficiency and accuracy.

Conclusively, the particle filter for self-localization has increased the localization accuracy significantly in contrast to using only the best match: from 0.84 m to 0.52 m in case of the small wheeled robot and from 11.55 m to 6.43 m in case of the quadrocopter. This corresponds to a reduction of the localization error of about 38.1 % and 44.3 %.

The computation time of the entire localization process comprises the time to extract the feature of the image and the comparison of this feature to all or a subset of features of the training images. Figure 4.13 depicts the feature extraction time in case of the small wheeled robot using various image features on different image resolutions. Scaling down the images speeds up the process especially in case of the more sophisticated algorithms WGOH and WGII. This is because both methods perform more complex computations on each pixel than simple histograms. The values that are denoted for the pixelwise image comparison are composed of the time to grab a frame, to convert it to grayscale and to resize it. These computations are performed in all methods, except the grayscale conversion in case of the color histograms, and are included in the measured computation times.

	WGOH (120×88)	PIC (15×11)
50 training images	3.76 s	2.89 s
100 training images	4.96 s	4.14 s
190 training images	7.13 s	6.56 s
particle filter ($m = 20$)	3.87 s	2.63 s
particle filter ($m = 40$)	4.81 s	3.84 s

Table 4.2: Computation times of the retrieval process on the small wheeled robot by using different numbers of training images with the best match and a particle filter on the 190 training images with 20 and 40 particles.

Table 4.2 denotes the computation times for the entire process. While the feature extraction plus feature matching of one test image to all 190 training images needs 7.13 s in case of WGOH at 120×88 pixels, it can be speeded up to 3.84 s by using the particle filter with $m = 40$ and the pixelwise image comparison on the 15×11 images. This corresponds to an acceleration of 46.1 %. Overall, the entire process is still quite slow, but we also have to keep in mind the strong limitations of minirobots.

In case of the quadrocopter, the computation time of the best match was 0.64 s using WGOH at full resolution and 0.40 s using the pixelwise image comparison on the smallest resolution. This signifies that by the use of tiny images, an acceleration of 37.9 % could be achieved.

4.6 Summary

In this chapter we investigated efficient visual self-localization techniques for resource-limited mobile robots. We conducted extensive experiments under highly diverse conditions: on two different robot platforms in indoor as well as outdoor environments. Our overall aim was to achieve high localization accuracy while reducing computational costs and storage requirements.

For this purpose, several developments and adaptations have been made. A practical solution to specify the direction of the viewpoints and thus to facilitate visual self-localization was presented: The incorporation of a compass as a lightweight and low-cost device showed to be an adequate choice for resource-limited wheeled robots. Despite a relatively high rotational control error of the robots due to noisy odometry readings and wheel slip, our overall system has shown to be robust enough to cope with the resulting transformations in the images.

Another contribution to the performance of the system was the decision to use global image features. An intensive comparison of various features revealed that weighted gradient orientation histograms (WGOH) lead to the best overall localization results. Additionally, we reduced the image size to tiny resolutions of 11×15 and 20×15 pixels and were still able to establish self-localization at a good localization accuracy. This reduction made it also feasible to develop a straightforward minimalistic pixelwise image comparison which surprisingly yielded comparable results. By furthermore adding a particle filter to reduce the feature comparison time, the entire localization process could be speeded up by roughly 40 % in both scenarios.

Finally, an enhanced robustness to localization failures was achieved by combining global image features with the probabilistic technique of a particle filter. Customized observation models have been implemented and tested in this context that in the end convinced by the provided localization accuracies. In this way, the overall localization error both of the small wheeled robot and the quadrocopter could be decreased considerably by at least 38 %.

For the discussion of these results, it should be mentioned that in applications where time is a limiting factor, the deployment of wheeled minirobots must be considered carefully. Both the computation times required for self-localizing the minirobot and the rotation towards a specified mapping direction may be demanding in time-constrained scenarios. A different approach to tackle self-localization at shorter computation times could be to send the image to a "base station" which calculates the position. Possibly, this base station could be a robot such as the service robot described in Chapter 2. However, apart from such a base station, this technique strongly requires unrestricted communication. Additionally, we recall that many applications do not require a robot to self-localize in real-time and therefore easily allow the usage of the proposed self-localization techniques.

Chapter 5

Area Coverage with a Heterogeneous Robot Team

Area coverage is an essential task for mobile robots that finds its relevance in various applications, e.g., floor cleaning, lawn mowing or harvesting. Generally, in area coverage, robots pass over the entire free space of a target area at least once.

In this chapter, we propose a novel method allowing multiple wheeled and resource-limited minirobots, which are referred to as *child robots*, to cover an area efficiently. They are assisted by a robot with state-of-the-art processing power and capable sensors, named *parent robot*. This structure traces back to our heterogeneous robot team presented in Chapter 2 and exploits the advantages of two distinct robot platforms. Intensive simulation experiments are conducted to validate the proposed method. The physical robot team itself will be deployed in the cooperative mapping of the next chapter that builds on this area coverage technique. Overall, the present chapter is an extended version of our related publication [102].

After giving an introduction and a taxonomy of area coverage algorithms in the first section, we review related work on deploying heterogeneous groups of robots and on coverage techniques in Section 5.2. Then, we introduce our coverage approach in Section 5.3 and explain it into more detail in Section 5.4. In Section 5.5 we describe our experiments and the corresponding results. Finally, we give a summary of the proposed strategy in Section 5.6.

5.1 Introduction

Motion planning algorithms address the task to plan collision-free motions for bodies from a start to a goal position among a collection of static obstacles. Many solutions have been proposed in the literature and have been reviewed in several surveys and books [49, 138, 207]. Yet, these conventional motion planning algorithms do not address applications in which coverage, i.e., the visitation of an entire area, is required. Therefore, coverage path planning algorithms have been developed to tackle robotic tasks such as floor cleaning [178], demining [10], lawn mowing [18], snow removal [100] or harvesting [173].

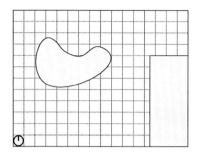 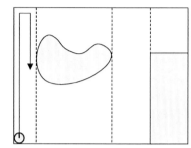

Figure 5.1: Different ways of decomposing an environment that is partially obstructed by objects. In the left image, approximate cellular decomposition is depicted, where a grid of equally sized cells was formed in the target area. Partially occupied cells are not accessible by the robot. Coverage algorithms here focus on determining the order in which these cells are visited. In the right image, an example of exact cellular decomposition is depicted with five cells of different sizes. These cells are assumed to be covered with simple back and forth motions of the robot.

In general, the problem of area coverage is related to the traveling salesman problem (TSP) and to its variant, the covering salesman problem (CSP). In the TSP, the shortest route for an agent has to be found, starting from a given city, visiting each of a specified group of cities, and then returning to the original point of departure [63]. In the CSP, which may be viewed as a generalization of the TSP, instead of visiting each city, an agent has to visit the neighborhood of each city within a certain distance [62]. In the theory of computational complexity, both TSP and CSP belong to the class of NP-hard problems. However, single-robot coverage has been shown to be solvable in polynomial time by making specific assumptions, e.g., building a regular grid of points that have to be visited as in the work of Gabriely and Rimon [81]. On the other hand, Zheng et al. [260] proved two natural variants of the multi-robot coverage problem to be NP-complete. Hence, we do not expect to be able to solve multi-robot coverage exactly in polynomial time, but believe that it can be tackled appropriately with a heuristic, as also stated by Zheng et al. [260].

One of the major benefits of recent coverage algorithms is that they commonly guarantee complete coverage, which former work often was not able to provide [48]. To achieve a form of provable guarantee, many coverage path planners either implicitly or explicitly use a cellular decomposition of the free space to achieve coverage. A cellular decomposition breaks down the target region into cells such that coverage is easy to determine. Provably complete coverage is then achieved if the robot is guaranteed to visit all cells in the work area.

In his survey on coverage path planning algorithms for robotics [48], Choset identified three types of decompositions: *approximate, semi-approximate* and *exact decomposition* methods. Whereas in approximate decomposition, the work area is approximated by a fine grid of cells in the free space, in exact decomposition, the set of differently sized regions entirely fills the free space. Figure 5.1 illustrates two types of area decomposition techniques. Choset also mentioned that, because complete coverage approaches require more sensory and computational power than simpler robots may possess, the ratio between costs and quality of the coverage may be better in randomized approaches. Low-cost cleaning robots and lawn mowers therefore often rely on such simpler methods.

An essential characteristic of coverage algorithms is whether information of the target region is available a priori, typically in form of a map. Such methods are referred to as *offline* in contrast to *online*, or *sensor-based*, coverage operations [48]. In online approaches, the environment has to be perceived by the robot's sensors during the operation. Therefore, these methods are more practicable in several applications, but it is harder to attain optimal coverage without knowing the entire workspace.

Finally, the time that is necessary to complete coverage is a central issue in most of the proposed algorithms. In this context, the use of multiple robots can obviously lead to high improvements and to shorter coverage times. Furthermore, multiple robots enhance the robustness, since failures of single robots do not necessarily terminate the entire operation.

Other topics related to area coverage are searching algorithms for static or moving targets [139] or robot exploration tasks [40, 108, 217]. However, searching tasks are different from area coverage in the sense that special attention may be paid to the trajectories of moving targets. Furthermore, the robot's task terminates as soon as a target is found, while in coverage, the task terminates only after the entire area has been visited. In the domain of explorations, robots usually have to explore an area by means of range sensors with a specific maximum distance depending on the sensor technology. This means that in contrast to area coverage, they do not have to pass over the entire area but might look over it from a certain distance in order to maximize knowledge about the environmental structure.

5.2 Related Work

While in Section 2.1, a taxonomy of multi-robot system architectures has been presented, the first part of this section provides a more practical overview of research focusing on the deployment of heterogeneous groups of mobile robots. Afterwards, we review related work on area coverage algorithms.

5.2.1 Heterogeneous Groups of Mobile Robots

One of the earliest research demonstrations of heterogeneity in mobile robot teams was presented in 1998 by Parker [181]. In the ALLIANCE architecture, she demonstrated the ability of robots to compensate for heterogeneity in task allocation and execution. The technique builds on the subsumption architecture [35] and adds behavior sets and motivations for achieving action selection without explicit negotiations among robots. Later on, the approach was extended to the L-ALLIANCE architecture [182] in which the performance of accomplishing a task influences on the robot's next task selection. Grabowski et al. [87, 168] developed centimeter-scaled robots, called millibots, which were configured by various modular components as cameras, sonar sensors and infrared sensors. The robots collaborated to explore and map unknown environment. Simmons et al. [218] proposed a general software architecture that permits heterogeneous robots to tightly work together with the purpose of achieving large-scale assembly and construction of objects.

Considerable progress in deploying teams of heterogeneous mobile robots was achieved by Howard et al. [108]. The objective of their work was to deploy a large number of simple robots as sensor nodes in an unexplored building by fewer, more capable robots and to map the building's interior in order to detect and track intruders. In the first part of their work, the more capable robots performed mapping and exploration by their laser range finders. Then, the robots provided navigation assistance to the simple robots, since they were not able to navigate independently. In this procedure described by Parker [185, 184], the simple robots are detected visually by means of unique fiducials. Overall, the deployed robot architecture is similar to ours. Yet, their simple robots are less integrated in the tasks than our child robots which directly participate in mapping and coverage.

Sukhatme et al. [223] conducted experiments by using a robotic helicopter and multiple wheeled ground robots. In their work, three case studies were demonstrated as examples of heterogeneous robot group control and coordination: marsupial-inspired payload deployment, cooperative localization, reconnaissance and surveillance. Further experiments in the coordination of aerial and ground robots were performed by Chaimowicz et al. [45], in which localization was provided to ground vehicles in areas where no GPS was available. Nowadays, an increasing number of researchers emphasize the coordination of heterogeneous groups of flying robots [175].

5.2.2 Coverage Algorithms

Motivated by the survey of Choset [48], we divide related work on area coverage algorithms into the following groups: randomized methods, approaches using approximate cellular decompositions, approaches using exact cellular decompositions and biologically inspired methods.

Randomized Approaches

Randomized approaches are usually not able to guarantee complete coverage, but exhibit several other advantages. Robots taking part in randomized coverage may not require expensive sensors for localization and do not need to spend computational resources for calculating their position and paths. Low-cost robots can be easily deployed in this way. Gage [82] analyzed the effectiveness of random versus coordinated search models. He further showed that random strategies become as effective as complete coverage when lots of inexpensive robots are used or the accuracy of the detector degrades [83]. MacKenzie and Balch [148] studied behavior-based vacuuming by combining a set of partially randomized behaviors. Recently, Palleja et al. [178] studied floor-cleaning coverage performances of commercially available domestic robots. Their results revealed that, as long as the exploratory algorithm of the cleaning robot was at least partially randomized, complete coverage was assured. Nevertheless, randomized approaches are likely to provide less efficiency in coverage, particularly when using multiple robots.

Approximate Cellular Decompositions

In approximate cellular decomposition, the work area is approximated by a fine grid of cells in the free space. This technique dates back to Moravec and Elfes [163], who modeled occupancy probabilities from sonar sensors as rasterized maps. Here, all cells are of equal size and shape and usually form a square (as an exception, Oh et al. [172] proposed a decomposition method using triangular cells). If the robot enters a cell, it is assumed that the cell has been fully covered. Accordingly, after all cells have been visited at least once by a robot, complete coverage has been attained. The size of the cells depends either on the size of the robot or its actuator, or on the resolution that is required by the operation.

By extending the distance transform path planning method of Jarvis [113], Zelinsky et al. [257] proposed a coverage path planning method. After all cells have been assigned values according to the original distance transform algorithm, instead of descending along the path of steepest descent to the goal, the robot follows the path of steepest ascent. It only moves into a cell which is closer to the goal if it has visited all neighboring cells that lie further away from the goal. A unique feature of this algorithm is that start and goal points can be specified.

A method that found large interest in the research community was the Spanning Tree Coverage (STC) approach by Gabriely and Rimon [81]. The original algorithm addresses single robot coverage and creates a spanning tree on a coarse grid of cells. A coarse grid cell consists of four cells lying in a square. On this coarse grid, a spanning tree is built such that all coarse grid cells are connected by the tree. By circumnavigating the tree, each of the fine grid cells is visited exactly once by the robot. Three STC versions were presented by Gabriely and Rimon [81]. In the first version, perfect a priori knowledge of the environment is assumed. An optimal coverage path can be constructed in time $O(N)$, where N is the number of cells comprising the area. The second version is of type online, where the robot uses its on-board sensors to construct a spanning tree while covering the work area. The online STC algorithm completes an optimal covering path in time $O(N)$, but requires $O(N)$ memory for its implementation. Finally, the third version of STC is "ant"-like. Again, the robot has no a priori knowledge of the environment, but it may have left pheromone-like markers during the coverage process. The ant-like STC algorithm runs in time $O(N)$, and requires only $O(1)$ memory. A general limitation of the STC approach is that coarse grid cells which are partially occluded by obstacles or bounds of the work area are fully discarded.

A similar approach to STC was proposed by Spires and Goldsmith [221]. In their work, robots had to traverse a Hilbert space-filling curve [201] whose range contains an entire two-dimensional square. However, this approach works only in obstacle-free areas which is a strong constraint.

Various multi-robot variants of the STC approach have been proposed so far [56, 94, 95, 97, 260]. Hazon and Kaminka [94] presented a family of algorithms called Multirobot Spanning-Tree Coverage (MSTC) that addresses robustness and efficiency. The *non-backtracking* version of the MSTC guarantees that the work area is completely covered in finite time, as long as at least a single robot is still able to move correctly. They analyzed the best- and worst-case completion times for the algorithm and found that in the worst-case, the coverage time for multiple robots is essentially equal to that of a single robot. Unfortunately, this scenario is quite common in coverage applications and occurs when all robots start near each other. *Backtracking* MSTC algorithms allow the robots to backtrack and thus let the robots visit cells at most twice. Although including redundancy by visiting cells more than once, the worst-case coverage time for the backtracking algorithms was shown to be half that of a single robot.

In a later extension of their work [95], Hazon et al. tackled not only the navigation techniques with a given spanning tree, but also the construction of the spanning trees to minimize coverage time. They expected that this problem is NP-hard, although it could not be proven yet. A polynomial-time tree construction algorithm was proposed that first builds N independent subtrees, where N is the number of robots deployed in coverage. Afterwards, the subtrees are connected to one single spanning tree by finding bridges that maximize parallel coverage. In their simulations, significantly better results were achieved than by building the spanning trees randomly. Later on, this technique was also applied to online coverage [96].

A different variant of the STC approach for multiple robots was suggested by Zheng et al. [260]. In their Multi-Robot Forest Coverage (MRFC), the work area is divided into a set of spanning trees, a spanning forest. The robots traverse them separately, i.e., each robot covers a single tree in the forest. They empirically showed improvements over the non-backtracking and backtracking versions of the

MSTC, but this comes at the cost that they can not guarantee robustness in case of robot failures. Furthermore, they assumed that multiple robots can occupy the same cell simultaneously and never block each other.

The work of Correll and Martinoli [55, 56, 57] is concerned with the boundary coverage problem for structures with identical elements which occurs in the inspection of the compressor section of a jet turbine engine. The boundary coverage problem was previously introduced by Easton and Burdick [69] and defines the problem of how a group of robots is able to inspect every point on the boundary of a two-dimensional environment. It was shown by Correll and Martinoli [55] that this problem is equivalent to the coverage of all vertices of a graph that further corresponds to the coverage of cells of a grid when the elements are aligned in a regular pattern (as is the case in the approaches above). Furthermore, they incorporated both probabilistic and deterministic elements into an STC algorithm [56]. Each of their robots constructed a spanning tree online and systematically explored it by a Depth-First-Search (DFS) algorithm. The system performance was analyzed under different levels of wheel slip. Rutishauer et al. [199] extended the approach by incorporating collaboration among the robots at different levels of communication ranges. Finally, Williams and Burdick [249] presented a graph-based approach for multi-robot boundary coverage and focused on the case where revisions of the original path planning are necessary due to changes in the environment or the robot setup.

Exact Cellular Decompositions

Exact cellular decomposition builds a set of non-overlapping regions, referred to as cells, whose union fills the entire free space of the target environment. In contrast to approximate cellular decomposition, the cells are of different size and shape. It is assumed that robots can cover the cells by simple back and forth motions. The path planning is thereby reduced to planning the transitions between the cells.

One popular cellular decomposition technique, which can yield a complete coverage path solution, is the trapezoidal decomposition by Latombe [137] and, equivalently under the name "slab", by Preparata and Shamos [188]. The trapezoidal decomposition assumes that a vertical line, termed a slice, sweeps left to right through a bounded environment with polygonal obstacles. Cells are formed via a sequence of open and close operations which occur when the slice intersects a vertex of a polygon.

At the border of cells, additional lengthwise motions are often necessary, which could be avoided by merging cells. To do so, *Boustrophedon cellular decomposition*[1]

---

[1]Boustrophedon literally means "the way of the ox". Typically, when an ox drags a plow in a field, it crosses the full length of the field in a straight line, turns around, and then traces a new straight line path adjacent to the previous one. By repeating this procedure, the ox is assumed to cover the entire field [50].

was proposed by Choset and Pignon [50] as an enhancement of the trapezoidal decomposition. It essentially merges cells lying at identical sides of objects. In the original version, the algorithm assumes a priori information about the environment and obstacle locations.

Acar and Choset [8] adapted the Boustrophedon decomposition for scenarios where the environment is not known in advance. A method was described to infer the location of "critical points" online from a simple sonar ring. Critical points were defined to be points where the connectivity of the slice changes, i.e., where new cells are created. An algorithm was developed which ensures a robot to encounter all critical points in the area, performing full decomposition and thus guaranteeing coverage of the unknown region. Later, Acar et al. [9] extended the method by introducing a hierarchical cell decomposition to ensure a robot with extended range sensor detectors to cover vast and open as well as narrow and cluttered spaces.

Huang [110] adopted the Boustrophedon method to further optimize coverage time. Instead of minimizing path length, his optimization criterion is the total number of turns required to cover all subregions. The underlying assumption is that turns are time-consuming because the robot has to decelerate, turn and accelerate again. Because the number of turns to cover a cell is proportional to the cell's width (perpendicular to the robot's motion direction), the goal is to choose a slice direction that minimizes the sum of the widths of all cells.

Another variant of a sensor-based coverage algorithm that uses a contact sensor was presented by Butler et al. [41]. It was later on extended to be used by multiple robots [42]. The algorithm termed CC_{RM} incrementally constructs a decomposition of the environment into cells with rectangular shapes. In the multi-robot algorithm, cooperation and coverage are algorithmically decoupled. This means that a coverage algorithm can be extended to a cooperative setting by adding an overseer algorithm which takes incoming data from other robots and integrates it into the cellular decomposition.

Latimer et al. [136] proposed a multi-robot algorithm in which robots move together in a formation covering the cells created by the decomposition. If a robot in the team encounters a critical point, the team divides to cover separate cells, until other critical points are detected that causes more subdivisions. Communication is restricted to members of a team and global localization is assumed for all robots.

Another set of multi-robot coverage algorithms based on the Boustrophedon decomposition was presented by Rekleitis et al. [191]. Algorithms were provided to make robots jointly cover a single cell and to allocate robots among multiple cells by using a greedy auction mechanism. Different communication ranges have been investigated. Acar et al. [10] also proposed a probabilistic coverage algorithm that is particularly suitable for demining: by learning the pattern in which the mines have been placed in the environment, the robot visits each mine location rather than covering the whole space exhaustively. Another domain where coverage path planning techniques are useful is in agriculture: Oksanen and Visala [173] provided some algorithms by taking into consideration specific agricultural aspects. Recently, Meuth et al. [152] proposed a set of memetic algorithms that decompose, allocate and optimize the exploration and coverage of a search area for multiple flying and ground vehicles.

Biologically Inspired Approaches

Some coverage approaches are inspired by biological principles that have been discovered in animal behaviors. Several of them make use of the principles of pheromones that are chemicals produced by insects such as ants. Pheromones can be used for various communication and coordination tasks. Wagner et al. [243] considered robots which leave chemical odor traces that evaporate with time. Three methods of covering a graph in a distributed fashion were presented, where smell traces are used. Later, both deterministic and probabilistic methods were discussed [244]. In recent work, the dirt on the floor was used as the main means of inter-robot communication [242].

Koenig et al. [129] studied search methods for ant robots that differ in how the markings in the terrain are updated. Bruemmer et al. [39] coordinated group behavior and promoted the emergence of swarm intelligence using social potential fields by means of attractive and repulsive fields emitted by each robot. Batalin and Sukhatme [27] proposed behavior-based algorithms for multi-robot coverage, where communication between the robots is limited to visual contact. They showed that the interaction between the robots is very simple, but their algorithm can not guarantee completeness. Menezes et al. [151] proposed a model of ant behavior to disperse robots in a random fashion. Finally, Spears et al. [220] created a physics-based algorithm for controlling a robot swarm. Their method is inspired by particle motion in gas, that is claimed to be an effective mechanism for expanding to cover complex environments.

5.3 Method Overview

In this section, a general introduction to our area coverage approach is given that uses a heterogeneous team of mobile robots. In terms of the previous section, we categorize our method to be of type offline, approximate cellular decomposition coverage. This means that the work area is known a priori and the free space is represented by a grid of cells that have to be visited. More specifically, for reasons of practicability, we establish a grid of vertices instead of cells. However, this is only a different way of illustration, as described in Section 5.4.2.

Our overall aim is to let the area be covered by the resource-limited child robots, since they can easily be deployed in larger numbers due to their low costs. However, since the child robots lack reliable sensors and possess only limited processing power, we do not expect that they have the ability to robustly navigate in the environment. This is a significant difference to the aforementioned multi-robot coverage algorithms. To let the child robots conduct the task regardless of these restrictions, we suggest the assistance of a more capable parent robot that is able to localize and robustly navigate by its rich sensors and state-of-the-art computational power. To enable a cooperation between the robots, the team has some specific capabilities, as already described in Chapter 2: the robots can communicate with each other and the parent robot is able to detect, track and thus teleoperate the child robots within its line-of-sight.

The first step of our approach is to define positions in the environment from which the parent robot can teleoperate the child robots to let them fulfill the coverage successfully. We recall that the child robots are assumed to only move safely within the pre-defined line-of-sight radius of the parent robot. The positions hence have to be chosen in such way that the entire area can be accessed by the child robots while simultaneously staying within the line-of-sight radius of the parent robot. Then, the order in which these positions are visited have to be determined. Finally, for each of the participating child robots, individual paths have to be planned to perform the actual coverage within line-of-sight of the parent robot's positions. Figure 5.2 depicts this basic approach.

Summarized, the structure of the algorithm is as follows.

1. From a given 2D occupancy grid map, vertices to be covered by the child robots are determined at regular distances.

2. A *parent roadmap graph* for the parent robot is built with *parent vertices* and *parent edges* such that all vertices lie in line-of-sight of at least one parent vertex. The parent edges determine the order of parent vertices that will be visited.

3. For each parent vertex, *child roadmap graphs* for all child robots are determined. In this way, vertices that lie in line-of-sight of a parent vertex will be covered.

4. When the coverage around a parent vertex is finished, the robots cooperatively move to the next one, until all parent vertices have been visited. This implies that all vertices from step (1) have been covered.

For conducting the actual coverage of the child robots within the line-of-sight of the parent vertices, we decided to extend the spanning tree-based coverage method for multiple robots by Hazon and Kaminka [97], since it exhibits an efficient coverage without visiting cells multiple times while providing a high robustness to robot failures.

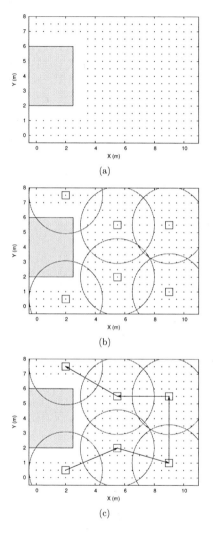

Figure 5.2: Basic coverage approach. (a) depicts the vertices to be covered in an environment that is partially obstructed by one object. In (b), parent vertices, depicted by small squares, have been chosen. The circles denote the corresponding line-of-sight radii of the parent robot standing on these positions. The order in which the parent vertices are visited is determined by parent edges (c).

5.4 Multi-Robot Coverage Considering Line-of-Sight Conditions

This section describes our coverage approach in detail. After providing a formalization of the algorithm, we describe how the grid of vertices is extracted and how the parent and child roadmap graphs are built. Finally, the way the robots transit from one parent vertex to the next is described.

5.4.1 Formalization

Let there be one parent robot and N child robots. From the given map, a grid of vertices $V = \{v^{(i)} \mid i = 1, ..., P\}$ to be covered by the child robots is extracted as described in Section 5.4.2.

A parent roadmap graph $G_p = (V_p, E_p)$ for the parent robot is defined that consists of Q parent vertices $V_p = \{v_p^{(t)} \mid t = 1, ..., Q\}$ and $Q - 1$ directed edges $E_p = \{(v_p^{(t)}, v_p^{(t+1)}) \mid t = 1, ..., Q-1\}$ with $V_p \subseteq V$. The parent robot is assumed to sequentially visit each parent vertex exactly once.

We further define the set of vertices in line-of-sight for a vertex v:

$$LOS(v) := \{w \mid w \in V \text{ with } ||w - v|| \leq r_{los} \text{ and}$$
$$v \text{ has unobstructed view towards } w\}, \tag{5.1}$$

where r_{los} is the line-of-sight radius of the parent robot. Since our parent robot is expected to have a height of approximately 1.5 m (cf. Section 2.2), it might be able to look over obstacles and get a view of vertices behind them. However, in reality this depends on the height of the obstacles which we do not necessarily know in advance. Hence, in Equation 5.1 we specified only those vertices to be in unobstructed view of a vertex v, which can be reached by a line, starting from v, without intersecting an obstacle. The obstacles are approximated by a number of squares of size $d \times d$ that are centered at the points of the grid, where d is the grid resolution.

To calculate the impact from the obstruction of vertices by objects, we also define for a vertex v the set

$$LOS_{freeView}(v) := \{w \mid w \in V \text{ with } ||w - v|| \leq r_{los}\}, \tag{5.2}$$

where obstacles do not obstruct direct views. We refer to this modification as *free view* in the following.

For every parent vertex $v_p^{(t)}$ at a stage t, a subset $C^{(t)}$ consisting of *child vertices* v_c can be computed, where

$$C^{(t)} = \left\{ v_c \in V \mid v_c \in LOS(v_p^{(t)}), v_c \notin \bigcup_{k=1}^{t-1} C^{(k)} \right\} \tag{5.3}$$

Assuming that for all $v \in V$ exists a parent vertex $v_p \in V_p$ such that $v \in LOS(v_p)$, it is guaranteed that

$$\dot{\bigcup_t} C^{(t)} = V \qquad (5.4)$$

For every $v_p^{(t)} \in V_p$ and for each child robot $j = 1, ..., N$ the child roadmap graphs $G_c^{(j)} = (V_c^{(j)}, E_c^{(j)})$ are constructed, such that $C^{(t)}$ is covered by the child robots. If $t < Q$, i.e., if there are still parent vertices to be visited, the robots move together to $v_p^{(t+1)}$, where new child roadmap graphs are computed as above. Consequently, if all $v \in V$ lie in line-of-sight of at least one $v_p \in V_p$ and if all $v_c \in C^{(t)}$ for all subsets $C^{(t)}$ are visited by the child roadmap graphs, it follows that V will be covered entirely.

Finally, some basic assumptions are made. We expect all robots to be able to move in the four directions up, down, left, right and we assume every vertex to be accessible in this way. As in most related approaches, diagonal movements of the robots are not permitted for reasons of simplicity. In the child roadmap graphs, each vertex is connected by l edges with $1 \leq l \leq 4$. In case of the parent roadmap graph, each vertex is connected by n edges with $1 \leq n \leq 2$. Furthermore, the graphs are entirely connected.

5.4.2 Extraction of the Grid of Vertices

Given the map of the environment that could be obtained by, e.g., a laser range finder, we extract a grid of vertices V that have to be covered. The distance d between the vertices is constant. Only vertices are chosen that lie in the center of a free cell of size $d \times d$. This procedure is equal to the construction of cells within free space of the room that was described in related work. Because of the construction of roadmap graphs, we considered our formulation to be more practical in this context.

5.4.3 Construction of a Parent Roadmap Graph

To construct G_p, V_p have to be chosen from V to fulfill Equation 5.4. Therefore, we pursue a technique on the set V that chooses vertices for V_p according to a greedy weighting strategy, called *boundary priority weighting*. The overall aim is to minimize $Q = |V_p|$. To evaluate this weighting, we additionally pursue a simpler method named *cardinality weighting*. In Figure 5.3, the two weighting methods are illustrated.

Boundary Priority Weighting

The boundary priority weighting starts choosing vertices near obstacles and boundaries of the environment, since in these zones, parent vertices are usually more difficult to extract. The area is then covered incrementally to avoid creating gaps between the parent vertices. More specifically, a weight is computed for every vertex that takes into account the number of vertices in line-of-sight and their corresponding number of direct neighbor vertices, while vertices with few neighbors are weighted higher. Direct neighbor vertices are the eight vertices lying around a vertex.

For all $v \in V$, the following steps are performed:

1. The set $LOS(v)$ is computed.

2. The number of direct neighbor vertices $n(v)$ is computed, where $1 \le n(v) \le 8$.

3. A specific weight $w_n(v) = (9 - n(v))^2$ is assigned to v.

After this, for all $v \in V$, a total score is computed:

$$s(v) = \sum_{l \in LOS(v)} w_n(l) \qquad (5.5)$$

The set V_p is then selected as follows. First, let $V_p := \emptyset$ and $M := \emptyset$. While $M \ne V$, the following steps are performed:

1. We define a set of neighboring vertices $N(V_p)$. In the first iteration of the algorithm, we set $N(V_p) = V$. In subsequent iterations, we define

$$N(V_p) := \{ v \in V \setminus V_p \mid \forall\, v_p \in V_p : ||v - v_p|| \le f \}, \qquad (5.6)$$

where f is approximately twice the maximal line-of-sight radius of the parent robot to subsequently cover the environment without creating gaps. The vertex $v \in N(V_p)$ with the highest score is then picked

$$\forall\, v' \in N(V_p) : s(v) \ge s(v') \qquad (5.7)$$

2. $V_p := V_p \cup \{v\}$

3. $M := M \cup LOS(v)$.

4. For all $v \in V$, $s(v)$ is updated.

After the set V_p is built, the order in which the vertices are visited can be interpreted as a traveling salesman problem (TSP). A standard heuristic such as a nearest neighbor algorithm can approximate its solution and compute E_p. To obtain the distances between the vertices, the Euclidean distance could be applied. However, in real-world environments, path lengths may be significantly larger due to obstruction by obstacles. Therefore we use the distance transform algorithm by Jarvis [113] to calculate the distance between every pair of $v_p^{(t)}$, which provides the corresponding shortest path.

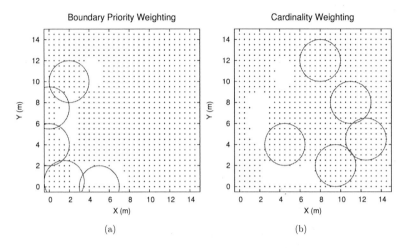

Figure 5.3: Illustration of the two weighting methods in an example environment. The line-of-sight radii of the five parent vertices which have been selected initially are depicted. In (a), the area is covered incrementally, starting from the boundary. In (b), parent vertices have been selected solely by their number of vertices in line-of-sight. Gaps might occur in this case.

Cardinality Weighting

To evaluate the proposed boundary priority weighting, we compare it to a simpler method, where for all $v \in V$, the total score $s(v)$ is simply computed by summing up the number of vertices lying in line-of-sight of it:

$$s(v) = | \, LOS(v) \, | \qquad (5.8)$$

The set V_p is then selected over the entire map as follows. Let $V_p := \emptyset$ and $M := \emptyset$. While $M \neq V$, the following steps are performed:

1. The vertex $v \in V$ is picked, ensuring

$$\forall \, v' \in V : s(v) \geq s(v') \qquad (5.9)$$

2. $V_p := V_p \cup \{v\}$

3. $M := M \cup LOS(v)$.

4. For all $v \in V$, $s(v)$ is updated.

The determination of E_p follows the procedure described in the boundary priority weighting.

Run-Time Complexity

Finally, we analyzed the running time of the suggested algorithms and came to a run-time complexity of $O(\frac{P^3}{Q})$, where $P = |V|$ and $Q = |V_p|$.

The following describes the derivation of the complexity. To calculate the total scores $s(v)$ for all $v \in V$, $LOS(v)$ has to be computed for each v. Let l be the maximal number of vertices lying in line-of-sight of a vertex. For all $a \in LOS_{freeView}(v)$, it has to be checked if

$$\exists\, b \in LOS_{freeView}(v) \text{ where } a \neq b \text{ and}$$
$$b \text{ obstructs the direct view from } v \text{ towards } a.$$

Thus, the computation of $s(v)$ for all $v \in V$ needs at most $O(P\,l^2)$ operations in the cardinality weighting. In the boundary priority weighting, $8\,P$ additional calculations are required due to the incorporation of the eight direct neighbor vertices for each $v \in V$. However, this is not a significant difference.

The selection of the Q parent vertices with maximal score $s(v)$ needs a run time of $O(Q\,P)$. With each selected parent vertex, scores are recalculated, which leads to an entire complexity of $O(Q(P + P\,l^2))$. l can be estimated to be $l \approx \frac{P}{Q}$, therefore we obtain a complexity of $O(Q(P + P\,(\frac{P}{Q})^2))$. Considering the highest order term only, this can be rewritten as $O(Q(P\,(\frac{P}{Q})^2))$ or $O(\frac{P^3}{Q})$.

Conclusively, our algorithm reveals a cubic complexity. The actual running time largely depends on l, which can be defined as $l = 4\,\pi\,(r_{los})^2$, where r_{los} is the line-of-sight radius of the parent robot. In reality, however, frequently less than $\frac{l}{4}$ vertices have to be checked for the determination whether the view on a vertex is obstructed or not.

5.4.4 Construction of Child Roadmap Graphs

After G_p has been created, for all $v_p^{(t)} \in V_p$ and for each child robot $j = 1, ..., N$, the roadmap graphs $G_c^{(j)} = (V_c^{(j)}, E_c^{(j)})$ are constructed such that the subset of child vertices $C^{(t)}$ will be covered.

To achieve efficient coverage paths for the child robots, we extend the spanning tree-based coverage method by Hazon and Kaminka [97]. Their algorithm creates spanning trees on a coarse grid of cells in the work area while considering the initial position of the robots. A *coarse grid cell* contains four child vertices that lie in a

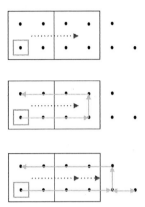

Figure 5.4: Creating a child roadmap graph. In the upper image, vertices to be covered and two coarse grid cells are depicted. The child robot is represented by a small square. The dotted arrows depict the tree that connects two coarse grid cells. In the middle image, the original version of the spanning tree technique is shown. In the lower image, our extended version is denoted. Continuous arrows represent the final roadmap graph of the child robot.

square. The challenge is to construct a spanning tree that minimizes the time to complete coverage. After the tree is built, the robots follow it around, creating a Hamiltonian cycle visiting all cells of the coarse grid.

Note that the algorithm of Hazon and Kaminka [97] requires that each coarse grid cell contains exactly four vertices to allow the circumnavigation of the subtrees. Individual vertices lying outside the coarse grid cells are generally not visited. However, since our aim is to make the algorithm applicable in real-world scenarios, we found this restriction too strong and enhanced the technique to also work with cells that lie outside the coarse grid. Figure 5.4 illustrates the original procedure and our extensions. The growing of the example tree starts in the coarse grid cell of the child robot. After the coarse grid cells are connected by the tree, three vertices remain unvisited in the original version. In the extended version, an additional *branch out*, i.e., enlargement, of the tree was done. After that, one vertex on the right is still not next to the tree. It can be accessed, but this requires the child robot to visit the vertex before twice and thus perform a redundant time step to return, which can not be avoided.

The creation of N roadmap graphs works as follows. First, a coarse grid is created over $C^{(t)}$. Then, the child robots are moved to the nearest free coarse grid

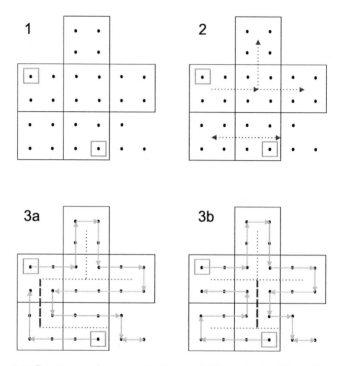

Figure 5.5: Creating roadmap graphs for two child robots. In (1), vertices to be covered and six coarse grid cells are shown. The child robots' positions are indicated by small squares. In (2), one subtree for each child robot has been built to cover the coarse grid cells and the remaining vertices. In (3a) and (3b), the subtrees have been connected by distinct bridges to a single spanning tree that is circumnavigated by the robots. The bridge of (3b) is finally chosen since it maximizes parallel coverage. Continuous arrows represent the final roadmap graphs of the child robots. Dotted arrows and lines depict the spanning trees.

cell that becomes the starting point for N separate subtrees. These subtrees grow independently for every child robot and will be connected later to a single spanning tree.

To create the subtrees, the following steps are performed for every child robot stepwise in parallel:

- For each unvisited neighboring coarse grid cell at the top of the tree, the Manhattan distance is computed to the current locations of all other child robots. The neighboring coarse grid cells are the cells lying on the left, on the right, above or below of the current coarse grid cell.

- The neighboring coarse grid cell that reveals the largest minimal distance to any other child robot is chosen to extend the subtree.

- If there is no unvisited neighboring coarse grid cell left at the top of the tree, the tree is *branched out*, i.e., it is extended beforehand in the tree.

- If the tree is entirely branched out on the coarse grid, it is further branched out on the original grid if possible, to also reach vertices which are not lying in a coarse grid cell.

Afterwards, all possible bridges between the N trees are searched to build one entire spanning tree. One advantage of a single tree instead of multiple ones is that it provides some robustness to robot failures: If a robot terminates to work, the remaining robots are able to continue the paths around the tree without replanning. For a specified number of cycles, the following steps are performed to find a good choice of $N - 1$ bridges that connect all subtrees.

- A valid set of bridges between the trees is created randomly such that one entire tree is assembled and all vertices remain accessible.

- The maximal distance between all child robots on the paths around the currently connected tree is then computed and saved.

The set of bridges that exhibits the smallest maximal distance between the child robots is taken to finally connect the subtrees. In this way, we make sure that the difference between the lengths of the robots' roadmaps on the tree is as small as possible, such that parallel coverage is enforced. Figure 5.5 illustrates an example where two child roadmap graphs are connected and circumnavigated. In Figure 5.10, the roadmap graphs have been constructed in the circular line-of-sight radius of the parent robot.

By circling the robots clockwise or counter-clockwise around the single spanning tree, we create the roadmap graphs $G_c^{(j)}$. Different methods were proposed by Hazon and Kaminka [97], where robots can follow the tree in one single direction only or also backtrack to the other direction if this saves coverage time. Here, we use the simple *non-backtracking multi-robot spanning-tree coverage (NB_MSTC)*, in which the robots start at their current positions and circle the tree until they reach the starting position of the next robot. To cover the parent vertex with a child robot, we assume that the parent robot leaves its vertex for one time step in a random direction and then returns.

Hazon and Kaminka [96] proved the tree creation algorithm to be of complexity $O(n^2 + N^k n)$, where n is the number of coarse grid cells and N is the number of subtrees (respectively robots). The exponent k denotes the number of cycles that are executed for finding suitable bridges. When N subtrees are created, the worst case in adding one coarse grid cell to a subtree is that the algorithm has to run over all present coarse grid cells in the subtree. This leads to a complexity of at most $O(n^2)$. In the second stage, when the trees have to be connected, N^k combinations of trees are examined, thus the relating complexity is $O(N^k n)$. The entire complexity reveals then to be $O(n^2 + N^k n)$.

5.4.5 Transitions Between Parent Vertices

After a parent vertex $v_p^{(t)}$ has been visited and $C^{(t)}$ covered, the team of robots has to move cooperatively to the next parent vertex. To do so, the common subset H of the two parent vertices is determined to switch from stage t to $t + 1$:

$$H := LOS(v_p^{(t)}) \cap LOS(v_p^{(t+1)}) \tag{5.10}$$

If the number of vertices in the subset is smaller than the number of deployed child robots, i.e., if $|H| < N$, a formation control task is performed that requires the robots to move together while maintaining certain relative positions. Since we want to build a "compact" formation to be able to pass narrow passages in the environment as well, we predefined following-positions around the parent robot for all child robots. After the coverage in line-of-sight has been finished, the child robots approach the parent robot to their nearest free following-positions. The parent robot is then able to guide the child robots in formation to the next parent vertex, where the robots continue coverage from their current positions. In the following, this behavior is called *formation transition*.

According to our experience, formation control of multiple robots is more prone to errors than steering the robots individually with a standing parent robot. If only one child robot fails to keep its position in the formation, either by a communication delay or wheel slip, the entire formation can fail. Furthermore, a moving parent robot means also that tracking losses get more probable due to a shaking camera. Therefore, we explicitly model the case when the line-of-sight radii of subsequent parent vertices overlap and $|H| \geq N$. In this case, the child robots can be moved in parallel to the nearest free vertex in H. After this, the parent robot moves

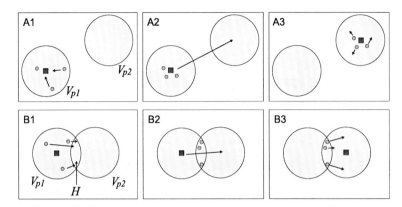

Figure 5.6: Transitions between the parent vertices. Three child robots and one parent robot, depicted as small circles and square, respectively, move from the line-of-sight radius of the first parent vertex V_{p1} to the radius of the next parent vertex V_{p2}. In case A, formation transition is illustrated: (A1) The child robots move to their following-positions. (A2) The team moves in formation to the next parent vertex. (A3) The parent robot reaches the parent vertex and the child robots continue coverage. In case B, direct transition is shown: (B1) The child robots move into the line-of-sight of the next parent vertex. (B2) The parent robot individually moves to the next parent vertex. (B3) The child robots continue coverage.

separately to $v_p^{(t+1)}$ and the child robots continue coverage. In the following, this behavior is termed *direct transition*. Figure 5.6 illustrates both types of transitions.

5.5 Experimental Results

To test our coverage approach, we performed various simulation experiments. We generated 2D grid maps of size 30 m×30 m corresponding to 3600 vertices. The distance d between the grid vertices was set 0.5 m. Then, we randomly added obstacles in different sizes. In general, the results of this section are averaged over ten cycles on randomly generated, different maps. To perform the boundary priority weighting to construct G_p, we set $f = 2\, r_{los}$, where r_{los} is the line-of-sight radius of the parent robot. The order of the parent vertices was determined by the TspSolve software of Hurwitz [6]. To find a set of suitable bridges for the purpose of constructing one single spanning tree, we conducted $2\,(N-1)$ cycles,

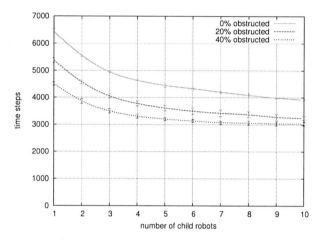

Figure 5.7: Required time steps for the coverage of different environments with a varying number of child robots.

where N is the number of deployed child robots, since this revealed to be a good compromise between computation time and solution quality. In our heterogeneous team of mobile robots (cf. Section 2.2), we found the parent robot to be able to detect the child robots robustly within $r_{los} = 2.0\,m$, which we chose as the default value in our experiments.

Figure 5.7 depicts the time steps required to cover differently obstructed environments with a varying number of child robots. r_{los} was set 2.0 m and the ratio of the obstructed area was set 0 %, 20 % and 40 % of the map. One time step comprises one movement of a robot by distance d in the direction up, down, left or right. Parallel movements of the robots are included in such time steps, i.e., they count only once.

Figure 5.7 shows that by deploying more child robots, the total coverage time significantly decreases. The minimal number of coverage time steps reached was 3920 (3242, 3037) at 0 % (20 %, 40 %) obstruction by employing the maximal number of 10 child robots, respectively. Table 5.1 shows the number of parent vertices that have been extracted. In obstructed environments, the number of parent vertices is smaller than in case of 0 % obstruction. Two contrasting factors may influence this: In obstructed environments, less vertices have to be covered, which obviously leads to a smaller number of parent vertices. On the other hand, obstacles disturb direct views and therefore also require more parent vertices.

To further investigate the results, we distinguish between *coverage time steps*

	0 % obstr.	20 % obstr.	40 % obstr.
$Q = |V_p|$	138.0	122.1±2.85	122.1±2.88

Table 5.1: Number of parent vertices Q at differently obstructed maps. Since at 0 % obstruction, the map is always equal, no standard deviation is given there.

that are performed by the child robots around the parent vertices in the "working phase", and the remaining *coordination time steps*. The coordination time steps include the time to pass from one parent vertex to the next one, either in formation transition or in direct transition. The formation transition further includes the time to initialize the formation. Figure 5.8 (a) depicts how the two phases are distributed. As expected, the coverage time steps decrease with the number of child robots. In parallel, the number of coordination time steps increases since a larger number of child robots require more time to be organized.

An essential parameter for performing the coverage in the team is the maximal line-of-sight radius r_{los} of the parent robot. Depending on the robot configuration, different line-of-sight radii are plausible. To investigate the influence of this parameter on coverage time, we chose different values for r_{los}, specifically 1.0 m, 2.0 m, 4.0 m and 8.0 m. Table 5.2 depicts the number of extracted parent vertices at different line-of-sight radii, where the boundary priority weighting was used. Figure 5.8 (b) illustrates the distribution of the time steps. By enlarging r_{los} up to 4.0 m, the total coverage time can be decreased. Overall, the coverage time steps at larger r_{los} only slightly reduce, while the coordination time steps drop significantly. This is because fewer parent vertices require fewer transitions between the vertices, which are time-consuming phases. However, a line-of-sight radius of 8.0 m can not further reduce the total time. This is because a larger line-of-sight radius also implies a drawback: The fragmentation of vertices gets larger, that is, during the coverage phase, vertices have to be crossed that have been visited in an earlier iteration. The minimal number of time steps was 5299 (3242, 2451, 2486) at a line-of-sight radius of 1.0 m (2.0 m, 4.0 m, 8.0 m) by deploying 10 child robots (except at $r_{los} = 1.0\,m$ by deploying 9 child robots). The maximal acceleration

	r_{los}			
	1.0 m	2.0 m	4.0 m	8.0 m
$Q = |V_p|$	365.0±6.7	122.1±2.9	46.8±2.5	28.0±2.4
Formation transitions	316.2±63.1	65.1±17.4	12.6±4.0	6.7±3.2

Table 5.2: Number of parent vertices $Q = |V_p|$ and number of formation transitions at different line-of-sight radii using the boundary priority weighting.

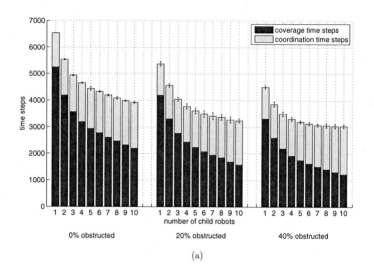

0% obstructed 20% obstructed 40% obstructed

(a)

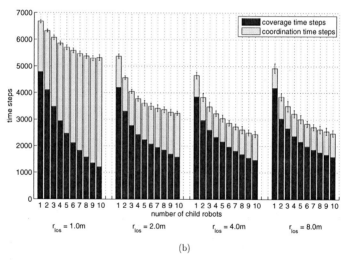

(b)

Figure 5.8: In (a), the distribution of total time steps for differently obstructed
environments is depicted. In (b), the distribution of total time steps at
maps with 20 % obstruction and different line-of-sight radii is shown.

	r_{los}			
	1.0 m	2.0 m	4.0 m	8.0 m
Boundary priority weighting	365.0±6.7	122.1±2.9	46.8±2.5	28.0±2.4
Cardinality weighting	369.9±6.2	135.1±3.6	53.3±3.0	29.9±3.3
Randomized	471.0±7.9	180.7±5.5	83.3±10.1	53.1±2.2
Free view	364.7±5.2	117.6±2.9	35.7±2.0	10.8±0.8

Table 5.3: Number of parent vertices $Q = |V_p|$ using various selection methods at different line-of-sight radii at 30 m×30 m maps with 20 % obstruction. In free view, boundary priority weighting is applied and obstacles do not obstruct direct views.

achieved by deploying multiple child robots in contrast to only a single one was 20.61 % (39.71 %, 47.29 %, 49.54 %) at a line-of-sight radius of 1.0 m (2.0 m, 4.0 m, 8.0 m), respectively.

Two types of transitions between the parent vertices have been proposed in Section 5.4.5: formation transitions and direct transitions. Table 5.2 depicts the number of transitions where formation transitions were used. Correspondingly, Figure 5.9 illustrates the percentage of transitions where formation transitions were conducted. In the rest of the transitions, direct transitions were used. In general, a larger number of child robots leads to a larger percentage of formation transitions. This is because direct transitions can only be performed if the number of common vertices lying in the line-of-sight of subsequent parent vertices is larger or equal the number of child robots ($|H| \geq N$). In addition, smaller line-of-sight radii also lead to a larger percentage of formation transitions, since overlaps are smaller and occur less often here.

To obtain the performances of the two transition variants, we measured the time steps required to execute them. We found that the averaged total number of time steps for executing formation transitions was by 10.35±1.98 % (16.57±5.11 %, 11.36±3.40 %) smaller than using them in combination with direct transitions at a line-of-sight radius of 2.0 m (4.0 m, 8.0 m). At $r_{los} = 1.0\,m$, no significant difference could be determined. However, as stated in Section 5.4.5, we wanted to reduce the number of formation transitions due to their relatively higher failure probability.

To evaluate our boundary priority weighting for selecting the parent vertices, we compared this method to the cardinality weighting as introduced in Section 5.4.3 and to two other variants. The resulting number of parent vertices $Q = |V_p|$ is depicted in Table 5.3. In general, less parent vertices are desirable since this leads to less coordination time steps. In the *randomized* version, the parent vertices were selected randomly from the entire map of vertices. In contrast to this randomization, the cardinality weighting reduces the number of parent vertices significantly by about 21.5 % (25.2 %, 36.0 %, 43.7 %) at a line-of-sight radius of 1.0 m (2.0 m, 4.0 m,

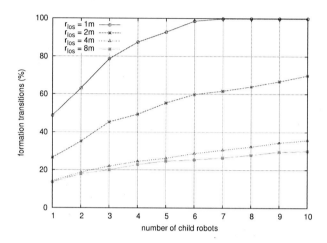

Figure 5.9: Percentage of formation transitions of all transitions between parent vertices at different line-of-sight radii. The rest of the transitions are direct transitions.

8.0 m). The incorporation of individual weights of vertices and of the neighborhood-based incremental coverage in the boundary priority weighting leads to a further reduction of the number of parent vertices of about 1.3 % (9.6 %, 12.2 %, 6.4 %). Obviously, the performance of the more sophisticated selection methods becomes valuable particularly with larger line-of-sight radii since overlaps occur more often here. Overall, by applying the boundary priority weighting, the number of parent vertices can be considerably reduced by 22.5 % (32.4 %, 43.8 %, 47.3 %) in contrast to the randomization at a line-of-sight radius of 1.0 m (2.0 m, 4.0 m, 8.0 m).

In Table 5.4, we measured the computation times of the entire coverage process for different line-of-sight radii on an Intel Dual-Core 4 with 3.0 GHz and 2.0 GB RAM on Ubuntu Linux 9.10. When $r_{los} = 8.0\,m$, the computationally most demanding factor is the geometrical determination of unobstructed vertices lying within the enhanced line-of-sight radius of the parent robot. The time to compute the child roadmap graphs also depends on the number of deployed child robots, where a larger number of child robots requires more time to find optimal bridges to connect their subtrees.

To further illustrate our approach, we provide a number of additional charts: Figure 5.10 exemplarily shows three child roadmap graphs that encircle a coverage spanning tree. Figures 5.11 and 5.12 depict an example environment where parent vertices and parent edges have been determined using different line-of-sight

	r_{los}			
	1.0 m	2.0 m	4.0 m	8.0 m
A	8.24±1.38 min	2.00±0.21 min	8.25±0.23 min	65.41±2.30 min
B	0.88±0.07 min	1.19±0.36 min	4.06±2.30 min	19.56±12.41 min
A+B	9.12±1.35 min	3.19±0.40 min	12.31±2.30 min	84.97±12.85 min

Table 5.4: Average computation times of the construction of the parent roadmap graph (A), the construction of the child roadmap graphs (B), and the entire process (A+B) for different line-of-sight radii.

radii. As described in Section 5.4, we assume that obstacles hide vertices from view by default since we do not know their heights in advance. To illustrate the impact of this constraint, we additionally provide maps where all vertices lying in line-of-sight of the parent vertices are assumed to remain visible, i.e., obstacles can be overlooked and do not obstruct direct views. The corresponding maps are shown in Figure 5.11 (d) and Figure 5.12 (b),(d). Particularly at larger line-of-sight radii, obstacles obstruct more often direct views. Therefore, free view reduces the corresponding number of parent vertices.

5.6 Summary

In this chapter, we presented a novel method to conduct area coverage by multiple mobile robots. Our work is one of the few examples that deploys a heterogeneous group of robots in this task. In our robot team, multiple resource-limited minirobots are assisted by a robot with state-of-the-art processing power and capable sensors. In this way, the minirobots are enabled to perform a task that they could hardly conduct alone.

As a drawback, the computation time of our method grows with larger line-of-sight radii of the parent robot. On the other hand, we do not require online processing in our method. In tasks that require the robots to cover the area multiple times, no new computation has to be performed.

We furthermore investigated to what extent the line-of-sight radius of the parent robot influences total coverage time. It resulted that a larger line-of-sight radius reduces coverage time as far as there is not too much fragmentation of the vertices within the line-of-sight radius. In general, larger line-of-sight radii reduce the number of parent vertices and thus also the time required for transitions between the parent vertices. Our greedy weighting technique for selecting the parent vertices additionally contributes to saving coverage time by considerably reducing the number of required parent vertices per map by about 23-47 %. Two different types of transitions between the parent vertices have been developed and examined that

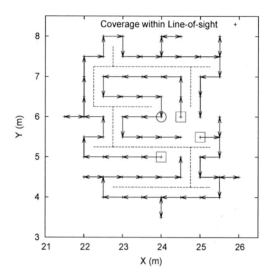

Figure 5.10: Coverage example within line-of-sight of the parent robot, which is illustrated as a small circle. Small rectangles depict the starting positions of three child robots and arrows indicate their roadmap graphs. Dotted lines denote the created spanning tree. We recall that the parent robot is assumed to leave its vertex for permitting a child robot to cover it and then returns.

are suitable for different situations: a technique that forms a formation between the robots which allows large distances to be passed and a direct transition method for neighboring parent vertices. Furthermore, we applied a popular spanning tree-based coverage technique for planning the paths of the child robots. This technique was chosen since it inherently provides efficiency and robustness to robot failures: as long as at least one child robot is able to move, coverage can be successfully terminated without replanning the robot paths. Moreover, we extended the spanning tree approach to also handle more realistic real-world scenarios.

Summarized, in contrast to related coverage algorithms, our robotic setup does not require that all participating robots are able to self-localize and robustly navigate. This benefit comes with the cost of additional time to coordinate the robots in the team. On the other side, by deploying multiple resource-limited robots, the total coverage time can be reduced by about 21-50 % in contrast to using only a single minirobot.

A natural extension to our work is the development of online coverage techniques for the heterogeneous robot team. In this case, the map of the environment would not be available a priori. However, this setup is likely to decrease coverage efficiency, since paths cannot be planned entirely and therefore easily lead to redundant visitation of parts of the area.

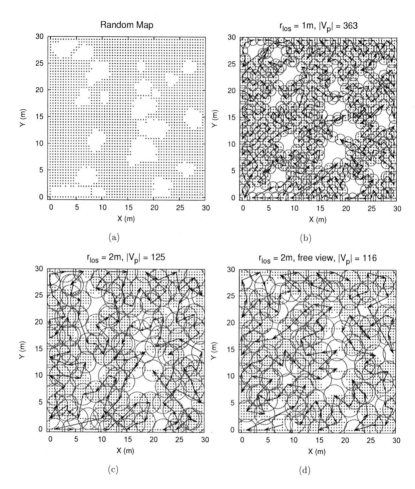

Figure 5.11: Coverage examples on a $30\,\mathrm{m} \times 30\,\mathrm{m}$ map with $20\,\%$ obstacles. Circles depict the line-of-sight radii of the parent vertices that were chosen by the boundary priority weighting. Arrows depict parent edges. (a) Randomly generated map of the environment. (b) Coverage where $r_{los} = 1\,m$. (c) Coverage where $r_{los} = 2\,m$. (d) Coverage where $r_{los} = 2\,m$ and free view, i.e., obstacles do not obstruct direct views.

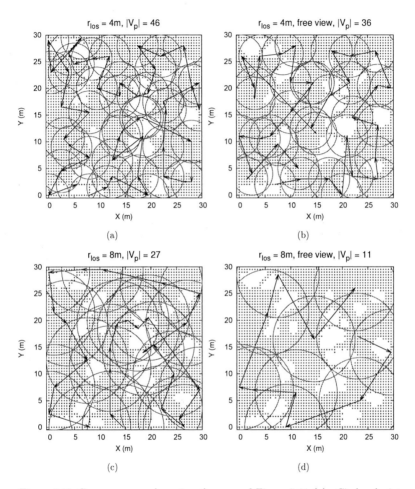

Figure 5.12: Coverage examples using the map of Figure 5.11 (a). Circles depict the line-of-sight radii of the parent vertices that are chosen by the boundary priority weighting. Arrows depict parent edges. (a) Coverage where $r_{los} = 4\,m$. (b) Coverage where $r_{los} = 4\,m$ and free view, i.e., obstacles do not obstruct direct views. (c) Coverage where $r_{los} = 8\,m$. (d) Coverage where $r_{los} = 8\,m$ and free view.

Chapter 6

Cooperative Visual Mapping with a Heterogeneous Robot Team

In Chapter 4 we discussed how resource-limited mobile robots can efficiently estimate their two-dimensional positions in the environment by means of their cameras. It was assumed that the robots were given a map of the environment in advance. However, manual mapping is time-consuming, particularly in large environments. In this chapter we therefore propose an autonomous mapping strategy for multiple resource-limited minirobots that traces back to our related publication [103]. It builds on the area coverage technique of the previous chapter and deploys a heterogeneous team of mobile robots.

At first, we give an introduction to state-of-the-art mapping approaches, which are frequently tackled jointly with self-localization, a process called *simultaneous localization and mapping (SLAM)*. After providing a definition and taxonomy of SLAM, in Section 6.2 we review related work on visual mapping techniques and methods which aim at deploying multiple robots in these tasks. In Section 6.3, we motivate and introduce our mapping approach and describe it in detail in Section 6.4. To validate the applicability of our technique, both real-world and simulation experiments have been conducted that are presented with the corresponding results in Section 6.5. Finally, Section 6.6 concludes this chapter with a short summary.

6.1 Introduction

Constructing maps is often referred to as simultaneous localization and mapping (SLAM) or concurrent mapping and localization (CML). SLAM is the process by which a mobile robot can build a map and at the same time use this map to deduce its own location. It is regarded as one of the most important problems towards the autonomy of mobile robots. Several tutorials and reviews have been published on this topic [22, 68, 231, 233].

The approach suggested in this chapter is an instance of *mapping with known poses*. It simplifies the mapping by assuming that the coordinates where the training images are grabbed are known. However, since this is a rather special case that arose from the use of resource-limited robots, we first give a definition and taxonomy of current simultaneous localization and mapping techniques designed for one or multiple robots.

The following describes a typical SLAM scenario: A mobile robot starts to move in an unknown environment from a location with known coordinates. Since its motion is uncertain due to control noise or exogenous effects such as wheel slip, the determination of its global position becomes more difficult over time. While moving around, the robot collects sensor data of the environment. The actual problem is now to build a map of the environment and to simultaneously estimate the position relative to it. The following terminology and taxonomy of probabilistic SLAM is partially borrowed from Thrun and Leonard [233].

Let m be a model of the world, i.e., a map, typically consisting of the locations of landmarks. *Landmarks* usually denote easily recognizable features of an environment. The path of the robot, that is, a sequence of positions that have been passed over by the robot by time t, is denoted by $X_t = \{x_0, x_1, ...x_t\}$. Let A_t be the corresponding sequence of measured motions of the robot by time t, typically odometry readings. The sequence of sensor observations taken until time t are further denoted by O_t.

Two variants of SLAM are distinguished in the literature: *full SLAM* and *online SLAM*. In full SLAM, the joint posterior over the entire path X_t and the map m are computed from the available data:

$$p(X_t, m \mid O_t, A_t) \tag{6.1}$$

In online SLAM, the posterior is estimated over the map and the current pose x_t instead of the entire path:

$$p(x_t, m \mid O_t, A_t) \tag{6.2}$$

Algorithms concerning online SLAM are usually incremental and are able to process one data item at a time.

One of the most challenging problems in SLAM is the *data association problem*. It can be described as finding the correct correspondences between old and newly acquired landmarks. Some SLAM algorithms make the assumption that the identity of landmarks is known, while others provide special mechanisms for estimating correspondences. Another challenge occurs if a robot reaches an already visited location. In *loop closing*, the robot has to recognize the position to be able to build a globally consistent map.

Generally, the literature distinguishes three main SLAM paradigms: *extended Kalman filter (EKF) SLAM*, *graph-based optimization SLAM* and *particle filter-based SLAM*. In EKF SLAM, a single state vector is used to estimate the locations of the robot and a set of features in the environment together with an associated error covariance matrix. This matrix represents the uncertainty of the estimates, including the correlations between the robot and feature state estimates. The technique was first introduced by Smith and Cheeseman [219] and Moutarlier and Chatila [164]. When the robot moves through the environment, the system state vector and the covariance matrix are updated using the extended Kalman filter [118]. As new features are observed, new states are added to the system state vector. Therefore, the size of the covariance matrix grows quadratically and solving the task becomes computationally demanding. To tackle this problem, several researchers have proposed extensions to EKF SLAM by decomposing the map into local submaps, for which covariances are maintained separately, e.g., Guivant and Nebot [89] or Williams et al. [250].

The second principal SLAM paradigm builds on a graph representation, where landmarks and robot locations can be thought of as nodes in a graph. Consecutive robot poses and links between poses and corresponding landmarks (that have been observed at those poses) are illustrated by arcs (similarly to edges). In this way, a sparse graph is formed. Arcs in the graph are soft constraints. By resolving them into a globally consistent estimate, the map and the robot path can be reconstructed. A number of efficient optimization techniques can be applied such as conjugate or stochastic gradient descent. In general, graph-based SLAM methods have the advantage to scale much better to high-dimensional maps than EKF SLAM. Lu and Milios [147] were historically the first in representing SLAM as a set of links between robot poses and in formulating a global optimization algorithm for generating a map from such constraints.

The third main SLAM paradigm makes use of particle filters which represent a posterior probability over a set of particles. Because of the high dimension of the state space in the SLAM problem, Rao-Blackwellization is applied. According to Thrun and Leonard [233], the formal introduction of the Rao-Blackwellized particle filter (RBPF) into the field of SLAM was done by Murphy and Russell [167], followed by the FastSLAM algorithm by Montemerlo et al. [161]. FastSLAM decomposes the SLAM problem into a robot localization problem and a collection of landmark estimation problems that are conditioned on the robot path estimate. Each particle of the RBPF represents a sample ("guess") of the robot's path and possesses a Kalman filter for each of the landmarks. In this way, FastSLAM can solve both the full and the online SLAM problem, since it updates only the most recent poses. Important extensions to FastSLAM have been provided, among others, by Hähnel et al. [91], who adopted the RBPF to handle grid-based maps, and Eliazar and Parr [72], whose DP-SLAM and tree structure provide efficient tree update methods for grid-based maps.

6.2 Related Work

After reviewing research that uses a camera as a sensor for environment mapping, exploration and mapping techniques designed for multiple mobile robots are presented. Note that visual SLAM is closely related to visual self-localization which was already surveyed in Chapter 3.

6.2.1 Visual Mapping

While laser range finders can be considered to be the sensor of first choice in solving the SLAM problem, there exist several reasons to deploy cameras in localization and mapping, as already mentioned in Chapter 1.

As an addition to laser-scanning SLAM, vision systems can be successfully applied for detecting loop closures, as Newman et al. [169] showed. When cameras are used as main sensors in SLAM, landmarks are usually extracted in the form of visual features of the images. The positions of landmarks are then tracked over time. Most of the approaches rely on SIFT features [146]. Se et al. [208, 209] extracted SIFT features from stereo images to build 3D maps. The robot's egomotion was then computed by incorporating odometry and tracking the image features. By matching small submaps of features, reliable global localization could be achieved. Elinas et al. [73] presented σSLAM, a variant of the RBPF for robots with stereo vision. Karlsson et al. [125] proposed the vSLAM algorithm in which the factorization of FastSLAM [161] is used and the robot path estimates are updated by a particle filter. Barfoot [25] estimated the 3D motion of a vehicle online in an outdoor environment by extracting SIFT features from stereo images and adapting FastSLAM to their scenario. Milford et al. [156] proposed an approach called RatSLAM that is biologically motivated by the hippocampus of rodents. They used a competitive attractor network to integrate odometric information with landmark sensing to form a consistent representation of the environment.

An appearance-based minimalistic SLAM solution was proposed by Rybski et al. [200], in which, instead of measuring relative positions to landmarks or features, visual signatures are taken as landmarks. An extended Kalman filter processes position measurements which are inferred from an image recognition algorithm together with odometry data. Similarly, Andreasson et al. [13] introduced a similarity-computation method for obtaining relative pose estimates of neighboring images. A globally consistent map was then obtained by applying a relaxation algorithm on the map. Tapus and Siegwart [229] built on their earlier technique of creating fingerprint sequences to describe locations. In their mapping, a heuristic detects whether the current location of the robot is similar to a previously mapped one or not. König et al. [130] combined a visual appearance-based approach with an RBPF and applied an adaptive sensor model that is estimated online. They further conducted graph matching to evaluate the likelihood of a given topological map.

A number of recent methods are inspired by bag-of-words image retrieval techniques developed in the computer vision community: Fraundorfer et al. [80] utilized an image retrieval system for topological localization, mapping and navigation. Cummins and Newman [61] set up a probabilistic approach called FAB-MAP that explicitly accounts for perceptual aliasing in the environment. Identical but indistinctive observations receive a low probability of having come from the same place. Konolige et al. [131] described a mapping approach where connections among different stereo views are formed by consistent geometric matching of their features. A vocabulary tree was used to propose candidate views. Recently, Erinc and Carpin [76] proposed a system for mapping and navigation with heterogeneous robots. Based on SIFT descriptors, they created an appearance-based model of the world and navigated using a method based on epipolar geometry.

6.2.2 Multi-Robot Mapping

Obviously, multiple mobile robots can map an environment potentially faster than a single robot. In multi-robot SLAM, the goal is to build a global joint map by multiple robots and to localize the robots simultaneously in the map. Inherently, besides the fusion of separately acquired maps, a joint mapping approach addresses also the coordination between the robots to make the mapping as efficient as possible. Mostly, this subtask is referred to as multi-robot exploration, where it is aimed at maximizing the explored area by spreading out the robots in an intelligent way.

In an early work by Cohen [51], a team of robots explored an environment using a random walk to detect a goal location. After the goal had been found, this information was communicated among the team members in direct line-of-sight towards a designated robot called navigator, which accordingly could navigate to that location. Billard et al. [32] introduced a probabilistic model that describes a multi-robot system and a dynamic environment for building a topography of the environment. The predictions of the model were compared to simulation results and physical experiments, showing a good correspondence.

Grabowski et al. [87] established mapping and exploration with a team of small mobile robots. A localization system based on sonar distance measurements between the robots determined the position of the robots in the group. Sensor data acquired from sonar, infrared, and camera sensors were combined by a team leader. Rekleitis et al. [192] improved mapping by reducing the odometry error during exploration. While one robot was exploring the environment, the second one tracked its movements and estimated its current position. Similarly, Rothermich et al. [195] proposed collaborative localization and mapping algorithms for a swarm of mobile robots. A distributed and dynamic task allocation algorithm enables some robots to move, while others act as landmarks in the environment. By measuring the empty space between the robots with an infrared-based communication system, a map of the free space was created.

The idea of frontier-based exploration by multiple robots was formally introduced by Yamauchi et al. [254]. In this work, the robots navigated towards the nearest boundaries between free and unexplored space, the frontiers, to gain new information. By introducing individual costs and utilities of reaching the frontier cells, Burgard et al. [40] extended the exploration approach. In their work, maps are exchanged and fused by the statistical framework of Thrun [230]. Specifically, Thrun combined an incremental maximum likelihood estimator with a posterior pose estimator to build consistent maps.

In a different work, Thrun and Liu [234] addressed the data association problem, which occurs when similar maps are created by multiple robots in case that the starting positions of the robots are not known. Howard et al. [109] developed a "manifold" map structure to provide self-consistency of maps. It takes the maps out of the plane onto a two-dimensional surface embedded in a higher-dimensional space. Later on, Howard [107] extended the RBPF to handle multi-robot SLAM problems in the case that the initial poses of the robots are known. If the poses are not known a priori, it was assumed that the robots can measure the pose relative to each other when they meet.

Recently, Gifford et al. [84] suggested an approach for autonomous mapping and exploration in unstructured environments such as deserts. After partitioning the area into subregions, multiple robots individually performed DP-SLAM [72] by low-cost range sensors and sent the created maps to a base in order to merge them. However, despite using resource-restricted platforms, their four-wheeled robots still provided considerably more processing power than our minirobots. Another multi-robot SLAM technique which builds a graph-like topological map was proposed by Chang et al. [46], in which vertices represent local metric maps and edges describe relative positions of adjacent local maps. Finally, Gil et al. [86] presented a multi-robot SLAM technique using vision. In their work, image features are extracted by a stereo camera and a joint map is built using an RBPF.

6.3 Method Overview

In the last section, a variety of mapping approaches has been reviewed. Although there exist solutions for mapping with limited sensing, e.g., by means of low-cost range finders [7, 30] or cameras [13, 200], such approaches aim at deploying only a single robot in this task. Building multiple maps, exchanging them between the robots, and finding correspondences between the maps become difficult tasks if only limited computational power is available. The requirement of an efficient exploration method further complicates the problem. The few works we are aware of which deployed multiple resource-limited robots in mapping [87, 195] created range-based maps by fusing distance measurements of range finders. In these approaches, robots stayed in line-of-sight of each other to act as landmarks for localization purposes. In the corresponding experiments, areas of only few square meters size were shown to be mapped.

In this work, we follow a different approach: We overcome the mentioned challenges by supporting the resource-limited *child robots* during the mapping phase by a more capable *parent robot* which is able to track and teleoperate the child robots within line-of-sight (cf. Section 2.2). Since the parent robot is able to robustly self-localize, it actively navigates the child robots to specified mapping positions. In this way, our approach becomes an instance of mapping with known poses.

In general, self-localization requires that a position has been visited and mapped before. Since our aim is to let the child robots navigate autonomously, e.g., in order to serve as mobile sensor nodes, we set the constraint that the target area has to be mapped entirely to ensure robust self-localization capabilities. To achieve this requirement, we build on multi-robot coverage as presented in Chapter 5. As in our offline coverage algorithm, we assume the existence of a prior map of the environment. We believe this to be a reasonable scenario, since the creation of laser grid maps[1] is common practice as shown in Section 6.2. Sometimes also blueprints of buildings are available.

Since our self-localization strategy of Chapter 4 yielded good results at low computational costs, we use it as a basis in this chapter. One key property of this technique was that, to specify the direction of viewpoints and thus facilitate self-localization, all test and training images have been grabbed pointing towards a single direction. The robot autonomously rotated towards this direction by means of its compass. In cooperative mapping, this requirement implies that no robot must stand in line-of-sight of another robot at the instance of grabbing an image.

As in coverage, the first step of our algorithm is to choose suitable positions for the parent robot such that all vertices are accessible by the child robots. Individual paths for the child robots are then calculated for each of these parent robot positions. The following enumeration briefly reviews the structure of our mapping procedure, incorporating the above constraints.

1. From a given 2D occupancy grid map, vertices at regular distances are determined that have to be visited by at least one child robot. At each of the vertices, an image has to be grabbed to establish self-localization capabilities.

2. A *parent roadmap graph* for the parent robot is built with *parent vertices* and *parent edges* such that all vertices lie in line-of-sight of at least one parent vertex. Since the images are grabbed in a single direction, we assign only vertices with unobstructed views, i.e., their views towards the direction of mapping are not obstructed by the parent robot standing on the parent vertex.

---

[1]Obviously, this range-sensor map does not permit the child robots to self-localize by means of their cameras. However, cameras are, as described in Chapter 1, advantageous sensors for resource-limited and small mobile robots.

The parent edges determine the order in which the parent vertices will be visited.

3. For each parent vertex, *child roadmap graphs* for all child robots are determined. In this way, vertices that lie in line-of-sight of a parent vertex will be visited, and images will be grabbed by the child robots at these points. Now, it must be taken care that no child robot stands in line-of-sight of another one when it is about grabbing an image.

4. When the mapping around a parent vertex is finished, the robots cooperatively move to the next one, until all parent vertices have been visited. This implies that all vertices from step (1) have been covered.

6.4 Multi-Robot Cooperative Mapping and Self-Localization

This section describes our cooperative mapping in detail. After extending the terminology of the last chapter, we describe the slightly changed construction of the parent roadmap graphs. Finally, we introduce a novel way to build the child roadmap graphs and describe the real-world implementation of the formation transitions.

6.4.1 Formalization

In the formalization of cooperative mapping, we make use of the terminology of multi-robot coverage presented in Section 5.4. We slightly modify the definition of $LOS(v)$ for a vertex v such that $v \notin LOS(v)$. In contrast to coverage, the child robots here do not visit the parent vertices V_p for reasons of simplicity, since our localization method is robust enough to handle these few missing reference positions. In addition, we define for a vertex v:

$$LOS_{mapping}(v) := \{w \mid w \in LOS(v) \text{ and } w \text{ has unobstructed} \atop \text{view towards the direction of mapping}\} \qquad (6.3)$$

We specify a direction of mapping α in that the test and training images will be grabbed. In contrast to the determination of $LOS(v)$, only vertices are chosen for $LOS_{mapping}(v)$ whose view towards α is not obstructed by the parent robot standing on the parent vertex. Formally, we specified a vertex w to have unobstructed view if the parent robot does not stand within the field of view of the camera. The opening angle of the camera β has its vertex in w and points towards α. The parent robot is located at the position of the parent vertex and is modeled by a square of size

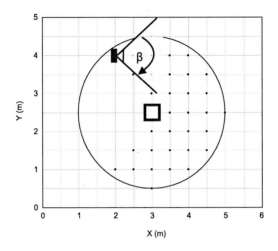

Figure 6.1: Determination of vertices that lie in line-of-sight of the parent robot and that have unobstructed view towards the direction of mapping α. Vertices lying in line-of-sight whose view is obstructed by the parent robot are not shown. The camera's opening angle β was set approximately 80° and $\alpha = right$. The parent robot is modeled by a square in the middle, its line-of-sight radius is denoted by a circle.

$d \times d$, where d is the grid resolution. Figure 6.1 provides an example, where vertices in line-of-sight are depicted whose view is not obstructed by the parent robot.

Concluding, for every parent vertex $v_p^{(t)}$ at a stage t, a subset $C^{(t)}$ consisting of *child vertices* v_c can be computed, where

$$C^{(t)} = \left\{ v_c \in V \mid v_c \in LOS_{mapping}(v_p^{(t)}), v_c \notin \bigcup_{k=1}^{t-1} C^{(k)} \right\} \qquad (6.4)$$

Assuming that for each vertex $v \in V \setminus V_p$ exists a parent vertex $v_p \in V_p$ such that $v \in LOS_{mapping}(v_p)$, it is guaranteed that

$$\dot{\bigcup_{t}} C^{(t)} = V \setminus V_p \qquad (6.5)$$

For every $v_p^{(t)} \in V_p$ and for each child robot $j = 1, ..., N$ the child roadmap graphs $G_c^{(j)} = (V_c^{(j)}, E_c^{(j)})$ are constructed such that $C^{(t)}$ is covered. Consequently, if all $v \in V \setminus V_p$ lie in line-of-sight of at least one $v_p \in V_p$ and if all $v_c \in C^{(t)}$ for all subsets $C^{(t)}$ are visited by the child roadmap graphs, it follows that $V \setminus V_p$ will be covered entirely.

Again, we make the assumption that all robots are able to move in the four directions up, down, left, right and that every vertex is accessible in this way. In the child roadmap graphs, each vertex is connected by l edges with $1 \leq l \leq 4$. In case of the parent roadmap graph, each vertex is connected by n edges with $1 \leq n \leq 2$. Furthermore, the graphs are entirely connected.

6.4.2 Construction of a Parent Roadmap Graph

After the grid of vertices has been extracted according to Section 5.4.2, the parent vertices and parent edges have to be determined. This procedure is analogous to the boundary priority weighting and the cardinality weighting of Section 5.4.3 with the exception that all sets $LOS(v)$ are replaced by the sets $LOS_{mapping}(v)$ since only vertices can be mapped which have unobstructed views towards the direction of mapping. Again, the parent edges can then be determined by a TSP-solving heuristic.

6.4.3 Construction of Child Roadmap Graphs

For the construction of the N child roadmap graphs, we now have to consider that no child robot obstructs the view of another one when it is grabbing an image. Therefore, in our algorithm the paths proceed towards the direction of mapping α and do not intersect themselves. Individual starting and goal points for the paths have to be chosen. Beginning from the vertices with the maximal distance to the direction of mapping, starting points are chosen while it is ensured that no starting point obstructs the view of another already chosen starting point. If more child robots are deployed than starting points can be found with the maximal distance to the direction of mapping, starting points are searched with a reduced distance to the direction of mapping.

Then, for each starting point, a goal point is chosen which lies at the minimal distance to the direction of mapping. Since, after the paths have been formed, they will be extended to cover the remaining neighbor vertices, the distances between the goal points are further maximized to partition the remaining vertices equally among the paths. If more child robots are deployed than goal points can be found at the minimal distance to the direction of mapping, goal points are searched with an increasing distance to the direction of mapping. By the distance transform algorithm of Jarvis [113], the paths can then be planned from the starting points to the goal points.

After the basic paths have been established, they are extended stepwise in parallel to cover all child vertices $C^{(t)}$ in the following way (see Figure 6.2):

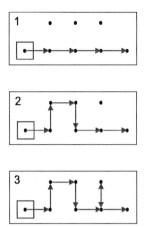

Figure 6.2: Creating a child roadmap graph for mapping. In (1), the original path is shown which proceeds in the direction of mapping α (towards right). In (2), the path is extended by the hilling procedure. (3) shows how an additional branch out of the path was performed to cover all vertices. The starting point of the child robot is illustrated as a square, arrows depict the planned robot's path.

- If there are two not yet visited neighbor vertices of the path, *hilling* is performed. In hilling, the path is winded and is extended onto two neighbor vertices. In this context, *neighbor vertices* are vertices lying next to the path and next to each other.

- If hilling cannot further be done, but if there are still remaining single unvisited neighbor vertices of the path, the path will be *branched out*. This means that the neighbor vertex is visited, but that the corresponding child robot has to perform a redundant time step to return.

In the end, all vertices are accessible by one of the child roadmap graphs. This procedure is similar to the tree building techniques developed by Hazon and Kaminka [97]. Figure 6.3 depicts an example of the construction of multiple child roadmap graphs where the direction of mapping $\alpha = right$. In this example, only two starting points were found at the maximal distance to the direction of mapping (at $x = 2.0\,m$). The third valid starting point that does not obstruct the view of the first ones was found at $(3.5, 2.5)$. Goal points were found at $x = 5.0\,m$ and $x = 4.5\,m$.

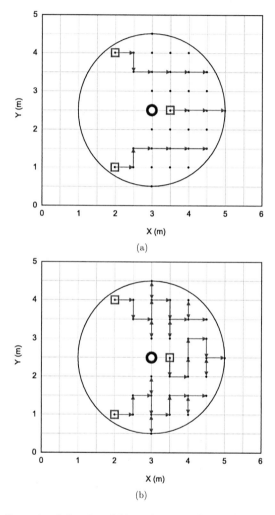

Figure 6.3: Example of planning child roadmap graphs on the map of Figure 6.1. Three child robots are deployed and $\alpha = right$. In (a), the original paths from the starting points to the goal points are depicted. In (b), the extended paths are shown where all vertices are covered. Rectangles depict the starting positions of the child robots and arrows their roadmap graphs. The position of the parent robot, that is, the parent vertex, is depicted by the small circle.

After the paths have been constructed and extended, they are followed by the child robots in parallel. Specifically, the robots move together, similarly to one vertical line (vertical to the direction of mapping), and proceed simultaneously towards the direction of mapping. In the example of Figure 6.3, this means that at the beginning of the mapping, only two robots are moving simultaneously while the one standing at $(3.5, 2.5)$ keeps waiting to avoid obstructing the view of the other robots. At the time the two robots have reached one vertex at $x = 3.5\,m$, all three robots move simultaneously towards the direction of mapping. In this way, all vertices lying in line-of-sight will be mapped entirely.

6.4.4 Transitions Between Parent Vertices

The transitions of the robot team between the parent vertices are performed as introduced in Section 5.4.5. The use of direct transitions or formation transitions depends on the number of vertices lying jointly within the line-of-sight radii of two subsequent parent vertices.

To successfully conduct the real-world formation transitions, we tested different following-behaviors. A chaining behavior where each child robot follows its predecessor by means of its cameras has the advantage that it is distributed, i.e., the computations are performed by all robots individually. However, we found this procedure not to be stable enough. The robots cut corners, sometimes overtook their leading robot and therefore easily collided with objects. Hence we decided to implement the following-behavior in a centralized way as it was also successfully done by Parker et al. [184, 185].

In our method, individual following-positions for the child robots are predefined in the neighborhood of the parent robot. The aim is to make the robots move closely together to allow also movements through narrow passages. However, a minimal distance of $0.6\,m$ between the child robots was required to avoid collisions and tracking errors. After the child robots initially moved to their nearest free following-positions, the parent robot starts to move towards the next parent vertex. During this procedure, it keeps track of the child robots' positions in the formation. By computing individual steering commands and sending them to the child robots, the robots keep moving in the formation. To calculate the steering commands, we adopted the behavior-based formation control technique for robot teams by Balch and Arkin [23]. In this technique, two behaviors control the speeds and headings of the robots according to their current and desired positions. See Figure 5.6 for an illustration of the formation transition procedure.

6.5 Experimental Results

Both simulation and real-world experiments were conducted to validate our mapping approach. After describing the experiments and their results in detail, we contrast the multi-robot coverage approach of the last chapter to the present cooperative mapping technique.

6.5.1 Simulation Results

We conducted various simulation experiments to test our approach. Analogously to Section 5.5, 2D grid maps of size 30 m×30 m were generated, which corresponds to 3600 vertices. The distance d between the grid vertices was set 0.5 m. Then, we randomly added obstacles of different sizes. To perform the boundary priority weighting to construct G_p, we set $f = 2\,r_{los}$, where r_{los} is the line-of-sight radius of the parent robot. To determine which vertices in line-of-sight of a parent vertex have unobstructed view towards the arbitrarily chosen direction of mapping $\alpha = west$, the opening angle of the child robots' cameras β was set to 80 °.

Figure 6.4 depicts the time that was required to map differently obstructed environments with a varying number of child robots. The results are averaged over ten cycles on randomly generated, different maps. r_{los} was set 2.0 m. The ratio of the obstructed area was set 0 %, 20 % and 40 % of the map. One time step comprises one movement of a robot by distance d in the direction up, down, left or right. We further distinguish between *mapping time steps* that are conducted by the child robots while grabbing images of the environment, and the remaining *coordination time steps* that are required to initialize and perform the transitions between the parent vertices and to guide the child robots to their mapping starting positions. Again, parallel movements of the robots are included in these time steps, i.e., they count only once.

Figure 6.4 shows that in obstructed environments the number of mapping time steps decreases due to the fact that fewer vertices have to be visited. By deploying a larger number of child robots, the mapping time steps generally decrease due to the parallel task execution. By contrast, the coordination time steps grow since more robots require more time to be coordinated, e.g., to form the formation transitions. The minimal total number of time steps was 4977 (4967, 4260) at 0 % (20 %, 40 %)

	0 % obstr.	20 % obstr.	40 % obstr.
$Q = |V_p|$	167.0	180.5±4.1	174.9±5.4

Table 6.1: Number of parent vertices $Q = |V_p|$ for differently obstructed maps. Since at 0% obstruction, the map is always equal, there is no standard deviation given.

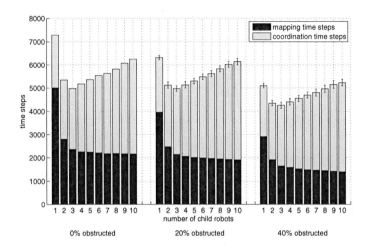

Figure 6.4: Distribution of total time steps for differently obstructed environments. Since at 0% obstruction the map is always equal, no standard deviations are given.

obstruction by employing three child robots, respectively. Table 6.1 lists the number of extracted parent vertices in the differently obstructed environments.

As in multi-robot coverage, depending on the robot configuration, different line-of-sight radii are plausible. In Table 6.2, the number of extracted parent vertices are shown at varying line-of-sight radii, specifically 1.0 m, 2.0 m, 4.0 m, and 8.0 m. Larger radii result in fewer parent vertices. In Figure 6.5, the time steps required at varying line-of-sight radii are depicted. By enlarging r_{los}, the total mapping time reduces. If r_{los} is small (1.0 m), multiple child robots can only slightly improve the

	r_{los}			
	1.0 m	2.0 m	4.0 m	8.0 m
$Q = |V_p|$	538.5±5.4	180.5±4.1	71.2±4.1	42.0±2.6
Formation transitions	403.1±154.7	54.8±18.1	18.4±5.6	9.5±3.6

Table 6.2: Number of parent vertices $Q = |V_p|$ and number of formation transitions at different line-of-sight radii, where the boundary priority weighting was used.

113

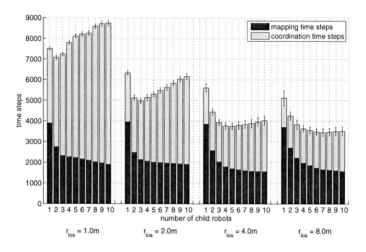

Figure 6.5: Distribution of total time steps at maps with 20 % obstruction and different line-of-sight radii.

total mapping time because of high coordination costs. The minimal number of time steps was 7082 (4967, 3752, 3446) at a line-of-sight radius of 1.0 m (2.0 m, 4.0 m, 8.0 m) by deploying 2 (3, 5, 7) child robots, respectively. This signifies that the larger the line-of-sight radius the larger is the number of child robots that can be deployed to reduce the total mapping time. In other words, a reasonable ratio between r_{los} and the number of child robots has to be chosen to minimize total mapping time.

In Section 5.4.5, direct transitions and formation transitions have been introduced to guide the robot team from one parent vertex to the next one. Table 6.2 shows the averaged number of formation transitions at different line-of-sight radii. In Figure 6.6, the distribution of the two types of transitions is depicted. As in coverage, a larger number of child robots leads to a higher percentage of formation transitions, because direct transitions can only be performed if the number of vertices lying in between two subsequent parent vertices is larger or equal the number of child robots ($|H| \geq N$). Smaller line-of-sight radii lead to a higher percentage of formation transitions due to fewer and smaller overlaps between the line-of-sight radii.

In Section 5.4.3, different methods for selecting the parent vertices were introduced. Table 6.3 shows the concluding number of extracted parent vertices. In the

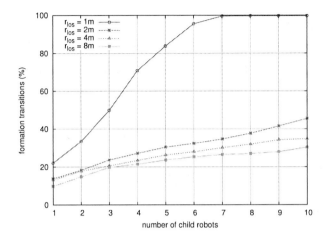

Figure 6.6: Percentage of formation transitions of all transitions between parent vertices at different line-of-sight radii. The rest of the transitions are direct transitions.

randomized version, the parent vertices are selected randomly of the entire map of vertices. In contrast to this randomization, our boundary priority weighting reduces the number of parent vertices about 20.3 % (39.5 %, 49.4 %, 53.5 %) at a line-of-sight radius of 1.0 m (2.0 m, 4.0 m, 8.0 m), respectively.

The measured computation times for the entire process at varying line-of-sight radii are shown in Table 6.4. Again, we conducted our computations on an Intel Dual-Core 4 with 3.0 GHz and 2.0 GB RAM on Ubuntu Linux 9.10. In contrast to multi-robot coverage, both the sets $LOS_{mapping}(v)$ and $LOS(v)$ have to be determined for a vertex v which makes it more computation-intensive, particularly at large line-of-sight radii.

6.5.2 Real-World Results

The real-world experiments were conducted with our heterogeneous team of mobile robots in an indoor environment of approximately $32 \, m^2$ (see Figure 6.7 (a)). The robot team consisted of one parent robot and three child robots since in our simulation, this configuration led to the minimal mapping time at $r_{los} = 2 \, m$. In total, 84 vertices were extracted from the occupancy grid map. The grid resolution d was set 0.5 m, β was set 80°, and the direction of mapping α was arbitrarily chosen west. As mentioned, the parent robot was able to robustly detect the child

	r_{los}			
	1.0 m	2.0 m	4.0 m	8.0 m
Boundary priority weighting	538.5±5.4	180.5±4.1	71.2±4.1	42.0±2.6
Cardinality weighting	540.5±6.2	188.2±3.0	79.9±5.3	44.1±3.1
Randomized	676.0±7.2	298.4±8.4	140.6±9.9	90.4±9.2
Free view	361.7±5.2	117.2±3.7	34.9±1.5	10.7±0.7

Table 6.3: Number of parent vertices $Q = |V_p|$ using various selection methods at different line-of-sight radii and 30 m×30 m maps with 20 % obstruction. In free view, boundary priority weighting is used and obstacles do not obstruct direct views.

	r_{los}			
	1.0 m	2.0 m	4.0 m	8.0 m
A	32.87±2.55 min	5.01±0.81 min	15.05±0.43 min	138.64±6.26 min
B	3.96±0.99 min	1.65±0.45 min	2.11±1.50 min	14.45±23.08 min
A+B	36.83±2.71 min	6.65±0.94 min	17.16±1.48 min	153.09±21.51 min

Table 6.4: Average computation times of the construction of the parent roadmap graph (A), the construction of the child roadmap graphs (B), and the entire computation (A+B) for different line-of-sight radii.

robots within a line-of-sight radius $r_{los} = 2.0\,m$. The number of determined parent vertices for this environment was seven.

Totally, we conducted the cooperative mapping ten times. The corresponding time to map the entire environment was 38.4±1.7 min. Since the child robots' movements are fairly imprecise, we steered the robots at 0.15 m per second only. This was also done to avoid child robot collisions and tracking losses which accordingly did not occur during the experiments. To approach a vertex we first made the child robots rotate towards the direction of the goal point by means of their compass and then perform a straight-line motion. Since compass data and robot rotations suffer from inaccuracies, we first let the child robots approach the goal point up to a distance of 0.2 m. If necessary, the child robots were then rotated again and steered straight-line directly towards the goal. This ensured reaching the vertices safely.

Figure 6.7 depicts the vertices extracted from the occupancy grid map and the actual paths of the child robots as tracked by the parent robot. The movements of the child robots were fairly imprecise: The position error between the child robots' desired mapping positions and their measured positions was 0.10±0.05 m. The rotational standard deviation between the robots' measured orientations and the

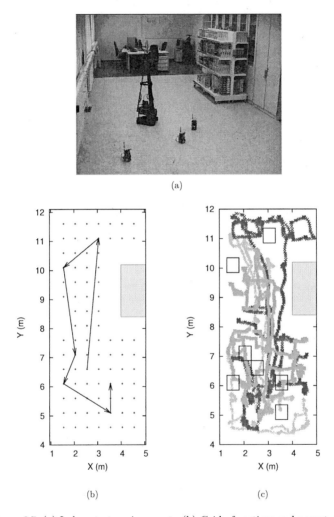

Figure 6.7: (a) Indoor test environment. (b) Grid of vertices and parent edges in the test environment. Arrows depict the parent edges and $\alpha = west = right$. (c) Paths of the three child robots as determined by the tracking system of the parent robot. Squares depict the positions of the parent vertices which did not have to be visited by the child robots.

desired mapping directions indicated by their compasses was $\sigma_{rot_{ct}} = 8.79 \pm 9.99°$.

In total, ten mapping datasets were created, consisting of 77 images each, since at the seven parent vertices, no images were taken. To test the localization accuracy of the images, we performed cross-validation on the ten datasets. For every image, the distance to the best match of the nine test datasets was determined. The best match was calculated as in Section 4.4.1. Two image features have been compared: weighted gradient orientation histograms (WGOH) at full image resolutions, since this feature led to best localization results in Chapter 4, and the straightforward pixelwise image comparison on the reduced image size, since this technique was one of the fastest ones of Chapter 4.

The mean localization error using WGOH with a resolution of 120×88 pixels was 0.43±0.59 m. By using the pixelwise image comparison at the reduced resolution of 11×15 pixels, we obtained a mean localization error of 0.54±0.79 m. Although the mapped area is of size 32 m², we believe that the provided mapping technique works also in larger environments, since the localization methods have shown a similar performance within a larger area of approximately 75 m² in Chapter 4. Furthermore, the additional use of a particle filter incorporates the robot's motion and sensor models and therefore helps to make the localization robust.

6.5.3 Comparison to Multi-Robot Coverage

The presented cooperative visual mapping built on the multi-robot coverage technique of Chapter 5. While in area coverage, the goal was to pass over the entire free space of an area, in our cooperative mapping, the direction of mapping had to be considered for path planning. The fact that neither the parent robot nor one of the child robots is allowed to stand in line-of-sight of another child robot at the instance of grabbing an image necessarily leads to a larger number of parent vertices and to more constrained path planning for the child robots.

By comparing our two methods, it becomes apparent that the cooperative mapping needs more time than the area coverage to be entirely executed. Figure 6.8 compares the required time steps at varying line-of-sight radii. In total, the minimal number of time steps to map the environment was 25.18 % (34.73 %, 34.67 %, 27.86 %) larger than to cover it, at a line-of-sight radius of 1.0 m (2.0 m, 4.0 m, 8.0 m), respectively. In coverage, an increasing number of child robots nearly always speeds up the coverage time. In mapping, the deployment of each additional child robot leads to significantly higher coordination costs, particularly at smaller line-of-sight radii. Hence, while in coverage the deployment of the largest investigated number of child robots (10) led to the minimal coverage time (except at $r_{los} = 1.0\,m$ the deployment of 9 child robots), in mapping, the deployment of more than 2 (3, 5, 7) child robots could not further reduce the mapping time, at a line-of-sight radius of 1.0 m (2.0 m, 4.0 m, 8.0 m), respectively.

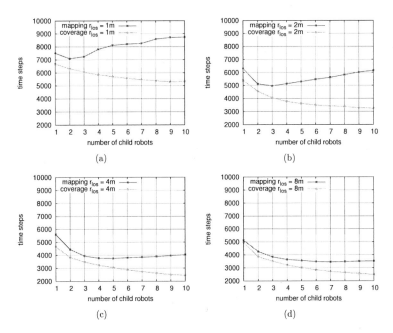

Figure 6.8: Comparison of coverage and mapping times required to fulfill the tasks. Varying line-of-sight radii are depicted.

6.6 Summary

In this chapter, a novel visual mapping strategy using a heterogeneous team of mobile robots was proposed that includes entire coverage of an area. This technique can be classified as a variant of mapping with known poses. By the assistance of a robot with capable sensors and state-of-the-art processing power, multiple resource-limited minirobots were enabled to map an environment entirely by means of their cameras. In the path planning of our technique, the direction of mapping was explicitly taken into consideration.

Real-world and simulation experiments attested the applicability of our method. In contrast to related approaches, our setup does not require all robots to possess accurate self-localization and navigation abilities.

After our mapping has been performed, no high-cost robot is required anymore: The recorded database of images can be sent to the minirobots, allowing them to act fully autonomously, e.g., as mobile sensor nodes. Furthermore, our approach is able to handle the challenging restrictions of small mobile robots, specifically the relatively large rotational control error and the imprecise movements.

As in multi-robot coverage, a possible further development is mapping in environments from which no range-based map is available a priori. However, this approach is likely to result in a less efficient mapping procedure due to redundant visits of parts of the area. Additionally, future work could investigate the influence of the distance between reference positions where training images are grabbed on the localization accuracy. Moreover, the discretization of the world could be performed differently from a regular grid: vertices may be chosen depending on environmental properties that reveal the self-similarity of the surrounding. In this way, mapping could be speeded up for larger environments.

Chapter 7
Summary and Outlook

In this thesis, we developed a number of novel strategies concerning the three robotic tasks self-localization, mapping and coverage. Apart from other studies, our techniques are particularly designed for resource-limited mobile robots. To conclude this thesis, we summarize the accomplished contributions in Section 7.1 and provide ideas for future research directions in Section 7.2.

7.1 Conclusions

In the first part of this thesis, we presented an overall localization system that enables resource-limited mobile robots to self-localize efficiently by means of their cameras. For wheeled minirobots, we newly developed a practical solution that integrates a compass to define the direction of viewpoints in which images are acquired. Basically, the overall localization system makes use of the fast computation of global image features and the robustness and strengths that come with the particle filter as a probabilistic method for self-localization. A variety of global image features were compared under different image resolutions to investigate localization accuracy and corresponding computation times. An interesting result is that tiny images at resolutions of 15×11 and 20×15 pixels still provided sufficient information to establish self-localization in medium-sized indoor and outdoor environments. A minimalistic pixelwise image comparison further yielded surprisingly robust results at minimal computational costs.

In the second part, we proposed a strategy that makes multiple resource-limited minirobots conduct area coverage in a robust and efficient way. A robotic team structure was developed in which a more capable service robot coordinates multiple minirobots that would hardly be able to perform such a complex task efficiently alone. In this team structure, we exploit the advantages of two different robot platforms: while the resource-limited minirobots are low-cost and therefore easily can be deployed in larger numbers, they benefit from the enhanced capabilities of a service robot that makes use of its accurate sensors and state-of-the-art computational power. We thoroughly investigated the impact that the physical team structure, e.g., the line-of-sight radius in which the minirobots can be teleoperated, has on the system's overall performance. In contrast to related work, we can relax the requirement that all robots must be able to autonomously self-localize and robustly navigate to take part in multi-robot coverage. Our work is one of the few examples that directly aims at deploying a heterogeneous group of robots in area coverage.

In the third part of this work, we utilized the proposed team structure to establish cooperative visual mapping. As a variant of mapping with known poses, the minirobots were enabled to map the entire free space of an environment visually. In real-world experiments it could be shown that the robots do not have to possess accurate navigation abilities to map the environment and to afterwards localize in it. The examination of the established visual datasets revealed a good localization accuracy in the medium-sized indoor environment, despite a relatively high position and rotation error of the robots. In this way, the proposed mapping strategy enables the minirobots to afterwards autonomously self-localize in the environment, e.g., as mobile sensor nodes, and perform a variety of tasks.

7.2 Future Work

Future work on visual self-localization could include the investigation of the proposed strategies in larger environments that moreover are dynamically changing. Several techniques have already been applied that address these challenges: the particle filter promises robust localization results also in large-scale environments due to its ability to form arbitrary probability distributions. Furthermore, several global image features specifically address dynamical changes occurring in parts of the image by using a subimage structure.

In the self-localization experiments, we also discovered the restrictions of our robot platforms concerning computation times. We recapitulate that from resource-limited mobile robots, we can not expect a performance that is equivalent to high-quality and expensive robotic platforms. However, since in many applications, real-time capabilities are not required, the advantages of resource-limited robots may outweigh the restrictions. Accordingly, in future work, one has to choose the type of robot carefully depending on the desired application.

As abovementioned, the suggested heterogeneous robot team structure has several advantages since it exploits the strengths of different robot platforms. However, the hierarchical structure comes with the cost of a centralization in form of the service robot. This implies that failures occurring in the central point may have influence on the entire robot team. For this reason, future work could include the cooperation between multiple heterogeneous teams of robots. To make the structure more flexible, a dynamic association of the minirobots to the service robots is imaginable. In this way, coverage or mapping could be partitioned more equally among the robots and thus could further contribute to saving operational time.

Moreover, the robotic team structure allows the execution of further tasks, e.g., indoor environmental monitoring in which the minirobots act as mobile sensor nodes. Information collected at the individual nodes could be passed to the service robot for evaluation purposes. By equipping the minirobots with related sensors, tasks as temperature monitoring in cold storage houses or surveillance missions in exhibition halls are reachable.

Bibliography

[1] Lego Mindstorm Robots. http://mindstorms.lego.com, December 2010.

[2] POB-Technology. http://www.pob-technology.com, December 2010.

[3] RoboCup Federation. http://www.robocup.org, December 2010.

[4] Federation of International Robot-soccer Association (FIRA). http://www.fira.net, December 2010.

[5] Inter-Process Communication (IPC). http://www.cs.cmu.edu/~ipc/, December 2010.

[6] The Stony Brook Algorithm Repository. http://www.cs.sunysb.edu/~algorith/implement/tsp/implement.shtml, December 2010.

[7] F. Abrate, B. Bona, and M. Indri. Experimental EKF-based SLAM for mini-rovers with IR sensors only. In *3rd European Conference on Mobile Robots (ECMR)*, pages 1–6, Freiburg, Germany, 2007.

[8] E. U. Acar and H. Choset. Sensor-based coverage of unknown environments: Incremental construction of morse decompositions. *The International Journal of Robotics Research*, 21(4):345–366, April 2002.

[9] E. U. Acar, H. Choset, and J. Y. Lee. Sensor-based coverage with extended range detectors. *IEEE Transactions on Robotics*, 22(1):189–198, February 2006.

[10] E. U. Acar, H. Choset, Y. Zhang, and M. Schervish. Path planning for robotic demining: Robust sensor-based coverage of unstructured environments and probabilistic methods. *The International Journal of Robotics Research*, 22(7–8):441–466, 2003.

[11] G. Adorni, M. Mordonini, S. Cagnoni, and A. Sgorbissa. Omnidirectional stereo systems for robot navigation. In *Conference on Computer Vision and Pattern Recognition Workshop (CVPRW)*, volume 7, pages 79–89, Madison, USA, 2003.

[12] A. Ali and J. K. Aggarwal. Segmentation and recognition of continuous human activity. In *Proceedings of the IEEE Workshop on Detection and Recognition of Events in Video*, pages 28–35, Vancouver, Canada, July 2001.

[13] H. Andreasson, T. Duckett, and J. A. Lilienthal. A minimalistic approach to appearance based visual SLAM. *IEEE Transactions on Robotics (Special issue on Visual SLAM)*, 24(4):991–1001, 2008.

[14] H. Andreasson, A. Treptow, and T. Duckett. Localization for mobile robots using panoramic vision, local features and particle filter. In *Proceedings of IEEE International Conference on Robotics and Automation (ICRA)*, pages 3348–3353, Barcelona, Spain, 2005.

[15] H. Andreasson, A. Treptow, and T. Duckett. Self-localization in non-stationary environments using omni-directional vision. *Robotics and Autonomous Systems*, 55(7):541–551, 2007.

[16] T. Arai, E. Pagello, and L. E. Parker. Guest editorial: Advances in multi-robot systems. *IEEE Transactions on Robotics and Automation*, 18(5):655–661, October 2002.

[17] S. Argamon-Engelson. Using image signatures for place recognition. *Pattern Recognition Letters*, 19(10):941–951, August 1998.

[18] E. M. Arkin, S. P. Fekete, and J. S. B. Mitchell. Approximation algorithms for lawn mowing and milling. *Computational Geometry: Theory and Applications*, 17(1–2):25–50, October 2000.

[19] M. Artač, M. Jogan, and A. Leonardis. Mobile robot localization using an incremental eigenspace model. In *Proceedings of IEEE International Conference on Robotics and Automation (ICRA)*, pages 1025–1030, Washington, USA, 2002.

[20] M. Artač and A. Leonardis. Outdoor mobile robot localisation using global and local features. In *Proceedings of the 9th Computer Vision Winter Workshop (CVWW)*, pages 175–184, Piran, Slovenia, 2004.

[21] M. S. Arulampalam, S. Maskell, N. Gordon, and T. Clapp. A tutorial on particle filters for online nonlinear/non-gaussian bayesian tracking. *IEEE Transactions on Signal Processing*, 50(2):174–188, February 2002.

[22] T. Bailey and H. Durrant-Whyte. Simultaneous localization and mapping (SLAM): Part II. *Robotics and Automation Magazine*, 13(3):108–117, 2006.

[23] T. Balch and R. Arkin. Behavior-based formation control for multi-robot teams. *IEEE Transactions on Robotics and Automation*, 14(6):926–939, December 1999.

[24] D. Ballard and C. Brown. *Computer Vision*. Prentice-Hall, 1982.

[25] T. D. Barfoot. Online visual motion estimation using FastSLAM with SIFT features. In *Proceedings of IEEE/RSJ International Conference on Intelligent Robots and Systems (IROS)*, pages 579–585, Edmonton, Canada, 2005.

[26] R. Basri and E. Rivlin. Localization and homing using combinations of model views. *Artificial Intelligence*, 78(1–2):327–354, 1995.

[27] M. A. Batalin and G. S. Sukhatme. Spreading out: A local approach to multi-robot coverage. In *Proceedings of the 6th International Symposium on Distributed Autonomous Robotics Systems (DARS)*, pages 373–382, Fukuoka, Japan, 2002.

[28] H. Bay, A. Ess, T. Tuytelaars, and L. van Gool. Speeded-up robust features (SURF). *Computer Vision and Image Understanding (CVIU)*, 110(3):346–359, June 2008.

[29] L. Bayindir and E. Şahin. A review of studies in swarm robotics. *Turkish Journal of Electrical Engineering & Computer Sciences*, 15(2):115–147, 2007.

[30] K. R. Beevers and W. H. Huang. SLAM with sparse sensing. In *Proceedings of IEEE International Conference on Robotics and Automation (ICRA)*, pages 2285–2290, Orlando, USA, 2006.

[31] J. Beis and D. Lowe. Shape indexing using approximate nearest-neighbour search in highdimensional spaces. In *Proceedings of the IEEE Conference on Computer Vision and Pattern Recognition (CVPR)*, pages 1000–1006, San Juan, Puerto Rico, June 1997.

[32] A. Billard, A. J. Ijspeert, and A. Martinoli. A multi-robot system for adaptive exploration of a fast-changing environment: Probabilistic modeling and experimental study. *Connection Science*, 11(3–4):359–379, 1999.

[33] D. M. Bradley, R. Patel, N. Vandapel, and S. M. Thayer. Real-time image-based topological localization in large outdoor environments. In *Proceedings of IEEE/RSJ International Conference on Intelligent Robots and Systems (IROS)*, pages 3670–3677, Edmonton, Canada, August 2005.

[34] T. J. Broida and R. Chellappa. Estimation of object motion parameters from noisy images. *IEEE Transactions on Pattern Analysis and Machine Intelligence*, 8(1):90–99, January 1986.

127

[35] R. A. Brooks. A robust layered control system for a mobile robot. *IEEE Journal of Robotics and Automation*, 2(1):14–23, March 1986.

[36] M. Brown and D. Lowe. Recognising panoramas. In *Proceedings of Ninth IEEE International Conference on Computer Vision (ICCV)*, pages 1218–1225, Nice, France, 2003.

[37] J. Bruce, T. Balch, and M. Veloso. Fast and inexpensive color image segmentation for interactive robots. In *Proceedings of IEEE/RSJ International Conference on Intelligent Robots and Systems (IROS)*, pages 2061–2066, Takamatsu, Japan, 2000.

[38] J. Bruce and M. Veloso. Fast and accurate vision-based pattern detection and identification. In *Proceedings of IEEE International Conference on Robotics and Automation (ICRA)*, pages 1277–1282, Taipei, Taiwan, 2003.

[39] D. J. Bruemmer, D. D. Dudenhoeffer, M. D. McKay, and M. O. Anderson. A robotic swarm for spill finding and perimeter formation. In *Spectrum*, Reno, USA, 2002.

[40] W. Burgard, M. Moors, C. Stachniss, and F. E. Schneider. Coordinated multi-robot exploration. *IEEE Transaction on Robotics*, 21(3):376–386, 2005.

[41] Z. J. Butler, A. A. Rizzi, and R. L. Hollis. Contact sensor-based coverage of rectilinear environments. In *Proceedings of IEEE International Symposium on Intelligent Control (ISIC)*, pages 266–271, Boston, USA, 1999.

[42] Z. J. Butler, A. A. Rizzi, and R. L. Hollis. Cooperative sensor-based coverage of rectilinear environments. In *Proceedings of IEEE International Conference on Robotics and Automation (ICRA)*, pages 2722–2727, San Francisco, USA, 2000.

[43] Y. U. Cao, A. S. Fukunaga, and A. B. Kahng. Cooperative mobile robotics: Antecedents and directions. *Autonomous Robots*, 4(1):1–23, 1997.

[44] C. Cauchois, E. Brassart, L. Delahoche, and T. Delhommelle. Reconstruction with the calibrated SYCLOP sensor. In *Proceedings of IEEE/RSJ International Conference on Intelligent Robots and Systems (IROS)*, pages 1493–1498, Takamatsu, Japan, October 2000.

[45] L. Chaimowicz, B. Grocholsky, J. F. Keller, V. Kumar, and C. J. Taylor. Experiments in multirobot air-ground coordination. In *Proceedings of IEEE International Conference on Robotics and Automation (ICRA)*, pages 4053–4058, New Orleans, USA, 2004.

[46] H. J. Chang, C. S. Lee, C. Y. Hu, and Y.-H. Lu. Multi-robot SLAM with topological/metric maps. In *Proceedings of IEEE/RSJ International Conference on Intelligent Robots and Systems (IROS)*, pages 1467–1472, San Diego, USA, 2007.

[47] T. Chen, H. Haussecker, A. Bovyrin, R. Belenov, K. Rodyushkin, A. Kuranov, and V. Eruhimov. Computer vision workload analysis: Case study of video surveillance systems. *Intel Technology Journal*, 9(2):109–118, May 2005.

[48] H. Choset. Coverage for robotics - a survey of recent results. *Annals of Mathematics and Artificial Intelligence*, 31(1-4):113–126, 2001.

[49] H. Choset, K. M. Lynch, S. Hutchinson, G. Kantor, W. Burgard, L. E. Kavraki, and S. Thrun. *Principles of Robot Motion: Theory, Algorithms, and Implementations*. MIT Press, Cambridge, USA, 2005.

[50] H. Choset and P. Pignon. Coverage path planning: The boustrophedon decomposition. In *Proceedings of International Conference on Field and Service Robotics (FSR)*, Canberra, Australia, December 1997.

[51] W. W. Cohen. Adaptive mapping and navigation by teams of simple robots. *Robotics and Autonomous Systems*, 18(4):411–434, 1996.

[52] D. Comaniciu and P. Meer. Mean shift: A robust approach toward feature space analysis. *IEEE Transactions on Pattern Analysis and Machine Intelligence*, 24(5):603–619, May 2002.

[53] D. Comaniciu, V. Ramesh, and P. Andmeer. Kernel-based object tracking. *IEEE Transactions on Pattern Analysis and Machine Intelligence*, 25(5):564–575, May 2003.

[54] D. Comaniciu, V. Ramesh, and P. Meer. Real-time tracking of non-rigid objects using mean shift. In *Proceedings of IEEE Conference on Computer Vision and Pattern Recognition (CVPR)*, pages 142–149, Hilton Head Island, USA, 2000.

[55] N. Correll and A. Martinoli. Comparing coordination schemes for miniature robotic swarms: A case study in boundary coverage of regular structures. In *Proceedings of the 10th International Symposium on Experimental Robotics (ISER)*, Rio de Janeiro, Brazil, July 2006.

[56] N. Correll and A. Martinoli. Robust distributed coverage using a swarm of miniature robots. In *Proceedings of IEEE International Conference on Robotics and Automation (ICRA)*, pages 397–384, Rome, Italy, 2007.

[57] N. Correll and A. Martinoli. Towards multi-robot inspection of industrial machinery - from distributed coverage algorithms to experiments with miniature robotic swarms. *IEEE Robotics and Automation Magazine*, 16(1):103–112, March 2008.

[58] I. J. Cox and J. J. Leonard. Modeling a dynamic environment using a bayesian multiple hypothesis approach. *Artificial Intelligence*, 66(2):311–344, 1994.

[59] E. Şahin and W. M. Spears, editors. *Swarm Robotics. SAB 2004 International Workshop*, Santa Monica, USA, 2004. Springer-Verlag.

[60] E. Şahin, W. M. Spears, and A. F. T. Winfield, editors. *Swarm Robotics. SAB 2006 International Workshop*, Rome, Italy, 2006. Springer-Verlag.

[61] M. Cummins and P. Newman. FAB-MAP: Probabilistic localization and mapping in the space of appearance. *The International Journal of Robotics Research*, 27(6):647–665, 2008.

[62] J. R. Current and D. A. Schilling. The covering salesman problem. *Transportation Science*, 23(3):208–213, 1989.

[63] G. B. Dantzig, R. Fulkerson, and S. M. Johnson. Solution of a large-scale traveling-salesman problem. *Journal of the Operations Research Society of America*, 2(4):393–410, 1954.

[64] A. Doucet, N. de Freitas, and N. Gordon. *Sequential Monte Carlo Methods in Practice*. Springer-Verlag, 2001.

[65] G. Dudek, M. R. M. Jenkin, E. Milios, and D. Wilkes. A taxonomy for multi-agent robotics. *Autonomous Robots*, 3(4):375–397, 1996.

[66] G. Dudek and C. Zhang. Vision-based robot localization without explicit object models. In *Proceedings of IEEE International Conference on Robotics and Automation (ICRA)*, pages 76–82, Minneapolis, USA, 1996.

[67] N. Duffy, J. Crowley, and G. Lacey. Object detection using colour. In *15th International Conference on Pattern Recognition (ICPR)*, pages 700–703, Barcelona, Spain, 2000.

[68] H. Durrant-Whyte and T. Bailey. Simultaneous localization and mapping (SLAM): Part I. *Robotics and Automation Magazine*, 13(2):99–110, 2006.

[69] K. Easton and J. Burdick. A coverage algorithm for multi-robot boundary inspection. In *Proceedings of IEEE International Conference on Robotics and Automation (ICRA)*, pages 727–734, Barcelona, Spain, 2005.

[70] G. Edwards, C. Taylor, and T. Cootes. Interpreting face images using active appearance models. In *International Conference on Face and Gesture Recognition (FG)*, pages 300–305, Nara, Japan, 1998.

[71] R. Egas, D. P. Huijsmans, M. S. Lew, and N. Sebe. Adapting k-d trees to visual retrieval. In *Proceedings of Third International Conference of Visual Information and Information Systems (VISUAL)*, pages 533–540, Amsterdam, The Netherlands, 1999.

[72] A. I. Eliazar and R. Parr. DP-SLAM 2.0. In *Proceedings of IEEE International Conference on Robotics and Automation (ICRA)*, pages 1314–1320, New Orleans, USA, 2004.

[73] P. Elinas, R. Sim, and J. J. Little. σSLAM: Stereo vision SLAM using the rao-blackwellised particle filter and a novel mixture proposal distribution. In *Proceedings of IEEE International Conference on Robotics and Automation (ICRA)*, pages 1564–1570, Orlando, USA, 2006.

[74] S. Erhard. Visuelle Selbstlokalisation mit einem Quadrocopter. Master's thesis, Department of Computer Science, University of Tübingen, Germany, July 2008.

[75] S. Erhard, K.-E. Wenzel, and A. Zell. Flyphone: Visual self-localisation using a mobile phone as onboard image processor on a quadrocopter. *Journal of Intelligent and Robotic Systems*, 57(1–4):451–465, 2009.

[76] G. Erinc and S. Carpin. Image-based mapping and navigation with heterogeneous robots. In *Proceedings of IEEE/RSJ International Conference on Intelligent Robots and Systems (IROS)*, pages 5807–5814, St. Louis, USA, October 2009.

[77] O. Faugeras. *Three-Dimensional Computer Vision: A Geometric Viewpoint.* MIT Press, Cambridge, USA, 1993.

[78] P. Fieguth and D. Terzopoulos. Color-based tracking of heads and other mobile objects at video. In *Proceedings of the IEEE Conference on Computer Vision and Pattern Recognition (CVPR)*, pages 21–27, San Juan, Puerto Rico, June 1997.

[79] D. Fox, W. Burgard, and S. Thrun. Markov localization for mobile robots in dynamic environments. *Journal of Artificial Intelligence Research*, 11:391–427, 1999.

[80] F. Fraundorfer, C. Engels, and D. Nistér. Topological mapping, localization and navigation using image collections. In *Proceedings of IEEE/RSJ*

International Conference on Intelligent Robots and Systems (IROS), pages 3872–3877, San Diego, USA, October 2007.

[81] Y. Gabriely and E. Rimon. Spanning-tree based coverage of continuous areas by a mobile robot. In *Annals of Mathematics and Artificial Intelligence*, volume 31, pages 77–98, 2001.

[82] D. W. Gage. Randomized search strategies with imperfect sensors. In *Proceedings of SPIE Mobile Robots VIII*, volume 2058, pages 270–279, Boston, USA, 1993.

[83] D. W. Gage. Many-robot MCM search systems. In *Proceedings of the Autonomous Vehicles in Mine Countermeasures Symposium*, pages 4–7, Monterey, USA, 1995.

[84] C. M. Gifford, R. Webb, J. Bley, D. Leung, M. Calnon, J. Makarewicz, B. Banz, and A. Agah. A novel low-cost, limited-resource approach to autonomous multi-robot exploration and mapping. *Robotics and Autonomous Systems*, 58(2):186–202, 2010.

[85] A. Gil, O. Mozos, M. Ballesta, and O. Reinoso. A comparative evaluation of interest point detectors and local descriptors for visual SLAM. *Machine Vision and Applications*, 21(6):905–920, 2010.

[86] A. Gil, O. Reinoso, M. Ballesta, and M. Julià. Multi-robot visual SLAM using a rao-blackwellized particle filter. *Robotics and Autonomous Systems*, 58(1):68–80, 2010.

[87] R. Grabowski, L. E. Navarro-Serment, C. Paredis, and P. Khosla. Heterogeneous teams of modular robots for mapping and exploration. *Autonomous Robots - Special Issue on Heterogeneous Multirobot Systems*, 8(3):271–298, June 2000.

[88] H.-M. Gross, A. König, C. Schröter, and H.-J. Böhme. Omnivision-based probabilistic self-localization for a mobile shopping assistant continued. In *Proceedings of IEEE/RSJ International Conference on Intelligent Robots and Systems (IROS)*, pages 1505–1511, Las Vegas, USA, 2003.

[89] J. Guivant and E. Nebot. Optimization of the simultaneous localization and map building algorithm for real time implementation. *IEEE Transactions on Robotics and Automation*, 17(3):242–257, 2001.

[90] D. Gurdan, J. Stumpf, M. Achtelik, K.-M. Doth, G. Hirzinger, and D. Rus. Energy-efficient autonomous four-rotor flying robot controlled at 1 kHz. In *Proceedings of IEEE International Conference on Robotics and Automation (ICRA)*, pages 361–366, Rome, Italy, April 2007.

[91] D. Hähnel, W. Burgard, D. Fox, and S. Thrun. A highly efficient FastSLAM algorithm for generating cyclic maps of large-scale environments from raw laser range measurements. In *Proceedings of IEEE/RSJ International Conference on Intelligent Robots and Systems (IROS)*, pages 206–211, Las Vegas, USA, 2003.

[92] C. Harris and M. Stephens. A combined corner and edge detection. In *Proceedings of the 4th Alvey Vision Conference*, pages 147–151, Manchester, UK, 1988.

[93] R. Hartley. Self-calibration from multiple views with a rotating camera. In *Proceedings of the 3rd European Conference on Computer Vision (ECCV)*, pages 471–478, Stockholm, Sweden, May 1994.

[94] N. Hazon and G. A. Kaminka. Redundancy, efficiency and robustness in multi-robot coverage. In *Proceedings of IEEE International Conference on Robotics and Automation (ICRA)*, pages 735–741, Barcelona, Spain, 2005.

[95] N. Hazon and G. A. Kaminka. Constructing spanning trees for efficient multi-robot coverage. In *Proceedings of IEEE International Conference on Robotics and Automation (ICRA)*, pages 1698–1703, Orlando, USA, 2006.

[96] N. Hazon and G. A. Kaminka. The giving tree: Constructing trees for efficient offline and online multi-robot coverage. *Annals of Mathematics and Artificial Intelligence*, 52(2–4):143–168, 2008.

[97] N. Hazon and G. A. Kaminka. On redundancy, efficiency, and robustness in coverage for multiple robots. *Robotics and Autonomous Systems*, 56(12):1102–1114, December 2008.

[98] P. Heinemann. *Cooperative Multi-Robot Soccer in a Highly Dynamic Environment*. PhD thesis, Department of Computer Science, University of Tübingen, Germany, 2007.

[99] P. Heinemann, J. Haase, and A. Zell. A combined monte-carlo localization and tracking algorithm for robocup. In *Proceedings of IEEE/RSJ International Conference on Intelligent Robots and Systems (IROS)*, pages 1535–1540, Beijing, China, 2006.

[100] M. Hess, M. Saska, and K. Schilling. Autonomous multi-vehicle formations for cooperative airfield snow shoveling. In *Proceedings of 3rd European Conference on Mobile Robots (ECMR)*, Freiburg, Germany, 2007.

[101] M. Hofmeister, S. Erhard, and A. Zell. Visual self-localization with tiny images. In *Autonome Mobile Systeme 2009 - 21. Fachgespräch*, pages 177–184, Karlsruhe, Germany, December 2009.

[102] M. Hofmeister and M. Kronfeld. Multi-robot coverage considering line-of-sight conditions. In *7th IFAC Symposium on Intelligent Autonomous Vehicles (IAV)*, Lecce, Italy, 2010.

[103] M. Hofmeister, M. Kronfeld, and A. Zell. Cooperative visual mapping in a heterogeneous team of mobile robots. In *Proceedings of IEEE International Conference on Robotics and Automation (ICRA)*, Shanghai, China, 2011.

[104] M. Hofmeister, M. Liebsch, and A. Zell. Visual self-localization for small mobile robots with weighted gradient orientation histograms. In *Proceedings of 40th International Symposium on Robotics (ISR)*, pages 87–91, Barcelona, Spain, 2009.

[105] M. Hofmeister, P. Vorst, and A. Zell. A comparison of efficient global image features for localizing small mobile robots. In *ISR/ROBOTIK 2010 (Proceedings of the joint conference of ISR 2010 (41st International Symposium on Robotics) and ROBOTIK 2010 (6th German Conference on Robotics))*, pages 143–150. VDE Verlag, June 2010.

[106] I. Horswill. Polly: A vision-based artificial agent. In *Proceedings of the 11th National Conference on Artificial Intelligence (AAAI)*, pages 824–829, Washington, USA, 1993.

[107] A. Howard. Multi-robot simultaneous localization and mapping using particle filters. *International Journal of Robotics Research*, 25(12):1243–1256, 2006.

[108] A. Howard, L. E. Parker, and G. S. Sukhatme. Experiments with a large heterogeneous mobile robot team: Exploration, mapping, deployment and detection. *International Journal of Robotics Research*, 25(5):431–447, May 2006.

[109] A. Howard, G. S. Sukhatme, and M. J. Matari. Multi-robot mapping using manifold representations. In *Proceedings of IEEE International Conference on Robotics and Automation (ICRA)*, pages 4198–4203, New Orleans, USA, April 2004.

[110] W. H. Huang. Optimal line-sweep-based decompositions for coverage algorithms. In *Proceedings of IEEE International Conference on Robotics and Automation (ICRA)*, pages 27–32, Seoul, Korea, 2001.

[111] D. P. Huttenlocher, J. J. Noh, and W. J. Rucklidge. Tracking non-rigid objects in complex scenes. In *Proceedings of 4th International Conference on Computer Vision (ICCV)*, pages 93–101, Berlin, Germany, May 1993.

[112] M. Isard and A. Blake. Condensation - conditional density propagation for visual tracking. *International Journal of Computer Vision*, 29(1):5–28, 1998.

[113] R. Jarvis. Robot path planning: Complexity, flexibility and application scope. In *Proceedings of the 2006 International Symposium on Practical Cognitive Agents and Robots (PCAR)*, pages 3–14, Perth, Australia, November 2006.

[114] H. Jegou, M. Douze, and C. Schmid. Hamming embedding and weak geometric consistency for large scale image search. In *Proceedings of the 10th European Conference on Computer Vision (ECCV)*, pages 304–317, Marseille, France, October 2008.

[115] M. Jogan, M. Artač, D. Skočaj, and A. Leonardis. A framework for robust and incremental self-localization of a mobile robot. In *Proceedings of 3rd International Conference on Computer Vision Systems (ICVS)*, pages 460–469, Graz, Austria, 2003.

[116] M. Jogan, A. Leonardis, H. Wildenauer, and H. Bischof. Mobile robot localization under varying illumination. In *Proceedings of 16th International Conference on Pattern Recognition (ICPR)*, pages 741–744, Quebec, Canada, 2002.

[117] P. Jonker, J. Caarls, and W. Bokhove. Fast and accurate robot vision for vision based motion. In *4th International Workshop on RoboCup*, pages 149–158, Melbourne, Australia, 2000.

[118] R. E. Kalman. A new approach to linear filtering and prediction problems. *Transactions of the ASME - Journal of Basic Engineering*, 82:35–45, 1960.

[119] S. B. Kang. Catadioptric self-calibration. In *Proceedings of the IEEE International Conference on Computer Vision and Pattern Recognition (CVPR)*, pages 201–207, Hilton Head, USA, June 2000.

[120] K. Kanjanawanishkul. *Coordinated Path Following Control and Formation Control of Mobile Robots*. PhD thesis, Department of Computer Science, University of Tübingen, Germany, 2010.

[121] K. Kanjanawanishkul, M. Hofmeister, and A. Zell. Coordinated path following for mobile robots. In *Autonome Mobile Systeme 2009 - 21. Fachgespräch*, pages 185–192, 2009.

[122] K. Kanjanawanishkul, M. Hofmeister, and A. Zell. Smooth reference tracking of a mobile robot using nonlinear model predictive control. In *4th European Conference on Mobile Robots (ECMR)*, pages 161–166, Mlini/Dubrovnik, Croatia, 2009.

[123] K. Kanjanawanishkul, M. Hofmeister, and A. Zell. Experiments on formation switching for mobile robots. In *IEEE/ASME International Conference on Advanced Intelligent Mechatronics (AIM)*, pages 323–328, Montréal, Canada, 2010.

[124] K. Kanjanawanishkul, M. Hofmeister, and A. Zell. Path following with an optimal forward velocity for a mobile robot. In *7th IFAC Symposium on Intelligent Autonomous Vehicles (IAV)*, Lecce, Italy, 2010.

[125] N. Karlsson, E. Di Bernardo, J. Ostrowski, L. Goncalves, P. Pirjanian, and M. E. Munich. The vSLAM algorithm for robust localization and mapping. In *Proceedings of IEEE International Conference on Robotics and Automation (ICRA)*, pages 24–29, Barcelona, Spain, April 2005.

[126] M. Kass, A. Witkin, and D. Terzopoulos. Snakes: Active contour models. *International Journal of Computer Vision*, 1(4):321–332, 1988.

[127] Y. Ke and R. Sukthankar. PCA-SIFT: A more distinctive representation for local image descriptors. In *Proceedings of the IEEE Conference on Computer Vision and Pattern Recognition (CVPR)*, pages 506–513, Washington, USA, 2004.

[128] B. Khoshnevis and G. Bekey. Centralized sensing and control of multiple mobile robots. *Computers and Industrial Engineering*, 35(3–4):503–506, December 1998.

[129] S. Koenig, B. Szymanski, and Y. Liu. Efficient and inefficient ant coverage methods. *Annals of Mathematics and Artificial Intelligence*, 31(1–4):41–76, 2001.

[130] A. König, J. Kessler, and H.-M. Gross. A graph matching technique for an appearance-based, visual SLAM-approach using rao-blackwellized particle filters. In *Proceedings of IEEE/RSJ International Conference on Intelligent Robots and Systems (IROS)*, pages 1576–1581, Nice, France, 2008.

[131] K. Konolige, J. Bowman, J. D. Chen, P. Mihelich, M. Calonder, V. Lepetit, and P. Fua. View-based maps. In *Proceedings of Robotics: Science and Systems*, Seattle, USA, 2009.

[132] D. Kortenkamp and T. Weymouth. Topological mapping for mobile robots using a combination of sonar and vision sensing. In *Proceedings of Twelfth National Conference on Artificial Intelligence (AAAI)*, pages 979–984, Seattle, USA, 1994.

[133] J. Kosecká and F. Li. Vision based topological markov localization. In *Proceedings of IEEE International Conference on Robotics and Automation (ICRA)*, pages 1481–1486, New Orleans, USA, 2004.

[134] A. Kuhn. Bildbasierte Objekterkennung von c't-Bots. Studienarbeit. Universität Tübingen, November 2008.

[135] P. Lamon, I. Nourbakhsh, B. Jensen, and R. Siegwart. Deriving and matching image fingerprint sequences for mobile robot localization. In *Proceedings of IEEE/RSJ International Conference on Intelligent Robots and Systems (IROS)*, pages 1609–1614, Seoul, Korea, 2001.

[136] D. Latimer, S. Srinivasa, and V. Lee-Shue. Towards sensor based coverage with robot teams. In *Proceedings of IEEE International Conference on Robotics and Automation (ICRA)*, pages 961–967, Washington, USA, 2002.

[137] J. C. Latombe. *Robot Motion Planning*. Kluwer Academic Publishers, Boston, USA, 1991.

[138] S. M. LaValle. *Planning Algorithms*. Cambridge University Press, Cambridge, USA, 2006.

[139] D. S. Lawrence. *Theory of Optimal Search*. Military Applications Society, 2004.

[140] J. J. Leonard and H. F. Durrant-Whyte. *Directed Sonar Sensing for Mobile Robot Navigation*. Springer-Verlag, 1992.

[141] L. Li, W. Huang, I. H. Gu, and Q. Tian. Foreground object detection from videos containing complex background. In *Proceedings of the 11th ACM International Conference on Multimedia*, pages 2–10, Berkeley, USA, 2003.

[142] M. Liebsch. Visuelle Selbstlokalisation von c't-Bots. Studienarbeit. Universität Tübingen, September 2008.

[143] J. S. Liu and R. Chen. Sequential monte carlo methods for dynamic systems. *Journal of the American Statistical Association*, 93(443):1032–1044, 1998.

[144] J. S. Liu, R. Chen, and T. Logvinenko. A theoretical framework for sequential importance sampling and resampling. Technical report, Stanford University, Department of Statistics, 2000.

[145] D. Lowe. Object recognition from local scale-invariant features. In *International Conference on Computer Vision (ICCV)*, pages 1150–1157, Corfu, Greece, 1999.

[146] D. Lowe. Distinctive image features from scale-invariant keypoints. In *International Journal of Computer Vision*, volume 60, pages 91–110, 2004.

[147] F. Lu and E. Milios. Globally consistent range scan alignment for environment mapping. *Autonomous Robot*, 4(4):333–349, 1997.

[148] D. MacKenzie and T. Balch. Making a clean sweep: Behavior-based vacuuming. In *AAAI Fall Symposium: Instantiating Real-world Agents*, Raleigh, USA, 1993.

[149] A. Martinoli and F. Mondada. Collective and cooperative group behaviours: Biologically inspired experiments in robotics. In *Proceedings of the 4th International Symposium on Experimental Robotics (ISER)*, pages 3–10, Stanford, USA, June 1995.

[150] S. J. Maybank and O. D. Faugeras. A theory of self-calibration of a moving camera. *International Journal of Computer Vision*, 8(2):123–152, August 1992.

[151] R. Menezes, F. Martins, F. E. Vieira, R. Silva, and M. Braga. A model for terrain coverage inspired by ant's alarm pheromones. In *Proceedings of the ACM Symposium on Applied Computing*, pages 728–732, Seoul, Korea, 2007.

[152] R. Meuth, E. Saad, D. Wunsch, and J. Vian. Memetic mission management. *IEEE Computational Intelligence Magazine*, 5(2):32–40, 2010.

[153] B. Micusik and T. Pajdla. Para-catadioptric camera auto-calibration from epipolar geometry. In *Proceedings of the Asian Conference on Computer Vision (ACCV)*, Jeju Island, Korea, January 2004.

[154] K. Mikolajczyk and C. Schmid. Indexing based on scale invariant interest points. In *Proceedings of Eighth IEEE International Conference on Computer Vision (ICCV)*, pages 525–531, Vancouver, Canada, 2001.

[155] K. Mikolajczyk and C. Schmid. A performance evaluation of local descriptors. *IEEE Transactions on Pattern Analysis and Machine Intelligence*, 27(10):1615–1630, October 2005.

[156] M. J. Milford, G. F. Wyeth, and D. Prasser. RatSLAM: A hippocampal model for simultaneous localization and mapping. In *Proceedings of IEEE International Conference on Robotics and Automation (ICRA)*, pages 403–408, New Orleans, USA, 2004.

[157] F. Mondada, M. Bonani, X. Raemy, J. Pugh, C. Cianci, A. Klaptocz, S. Magnenat, J.-C. Zufferey, D. Floreano, and A. Martinoli. The e-puck, a robot designed for education in engineering. In *Proceedings of the 9th Conference on*

Autonomous Robot Systems and Competitions, pages 59–65, Castelo Branco, Portugal, May 2009.

[158] F. Mondada, E. Franzi, and A. Guignard. The development of Khepera. In *Proceedings of the First International Khepera Workshop*, pages 7–14, Paderborn, Germany, 1999.

[159] F. Mondada, A. Guignard, M. Bonani, D. Bar, M. Lauria, and D. Floreano. SWARM-BOT: From concept to implementation. In *Proceedings of IEEE/RSJ International Conference on Intelligent Robots and Systems (IROS)*, volume 2, pages 1626–1631, Las Vegas, USA, 2003.

[160] M. Montemerlo, N. Roy, and S. Thrun. Perspectives on standardization in mobile robot programming : The carnegie mellon navigation (CARMEN) toolkit. In *Proceedings of IEEE/RSJ International Conference on Intelligent Robots and Systems (IROS)*, pages 2436–2441, Las Vegas, USA, 2003.

[161] M. Montemerlo, S. Thrun, D. Koller, and B. Wegbreit. FastSLAM: A factored solution to the simultaneous localization and mapping problem. In *Proceedings of the AAAI National Conference on Artificial Intelligence*, pages 593–598, Edmonton, Canada, 2002.

[162] H. Moravec. Visual mapping by a robot rover. In *Proceedings of the 6th International Joint Conference on Artificial Intelligence (IJCAI)*, pages 599–601, Tokyo, Japan, August 1979.

[163] H. P. Moravec and E. Elfes. High resolution maps from wide angle sonar. In *Proceedings of IEEE International Conference on Robotics and Automation (ICRA)*, pages 116–121, St. Louis, USA, 1985.

[164] P. Moutarlier and R. Chatila. An experimental system for incremental environment modelling by an autonomous mobile robot. In *First International Symposium on Experimental Robotics*, pages 327–346, Montréal, Canada, 1989.

[165] K. Murase, K. Sekiyama, N. Kubota, T. Naniwa, and J. Sitte, editors. *Proceedings of the 3rd International Symposium on Autonomous Minirobots for Research and Edutainment (AMiRE)*, Awara-Spa, Japan, 2005. Springer-Verlag.

[166] A. C. Murillo and J. Kosecká. Experiments in place recognition using gist panoramas. In *IEEE 12th International Conference on Computer Vision Workshops (ICCV Workshops)*, pages 2196–2203, Kyoto, Japan, 2009.

[167] K. Murphy and S. Russell. *Sequential Monte Carlo Methods in Practice*, chapter Rao-Blackwellized Particle Filtering for Dynamic Bayesian Networks, pages 499–516. Springer-Verlag, 2001.

[168] L. E. Navarro-Serment, R. Grabowski, C. J. Paredis, and P. K. Khosla. Millibots - the development of a framework and algorithms for a distributed heterogeneous robot team. *IEEE Robotics and Automation Magazine*, 9(4):31–40, December 2002.

[169] P. Newman, D. Cole, and K. Ho. Outdoor SLAM using visual appearance and laser ranging. In *Proceedings of IEEE International Conference on Robotics and Automation (ICRA)*, pages 1180–1187, Orlando, USA, 2006.

[170] S. Nouyan and M. Dorigo. Chain based path formation in swarms of robots. In *Proceedings of ANTS2006. 5th International Workshop on Ant Colony Optimization and Swarm Intelligence*, pages 120–131, Berlin, Germany, 2006.

[171] K. Nummiaro, E. Koller-Meier, and L. Van Gool. A color-based particle filter. In *First International Workshop on Generative-Model-Based Vision*, pages 53–60, Copenhagen, Denmark, 2002.

[172] J. S. Oh, Y. H. Choi, J. B. Park, and Y. F. Zheng. Complete coverage navigation of cleaning robots using triangular-cell-based map. *IEEE Transactions on Industrial Electronics*, 51(3):718–726, 2004.

[173] T. Oksanen and A. Visala. Coverage path planning algorithms for agricultural field machines. *Journal of Field Robotics*, 26(8):651–668, 2009.

[174] A. Oliva and A. Torralba. Building the gist of a scene: The role of global image features in recognition. *Visual Perception, Progress in Brain Research*, 155:23–36, 2006.

[175] A. Ollero and I. Maza, editors. *Multiple Heterogeneous Unmanned Aerial Vehicles*. Springer-Verlag, 2007.

[176] C. F. Olson. Probabilistic self-localization for mobile robots. *IEEE Transactions on Robotics and Automation*, 16(1):55–66, 2000.

[177] L. Paletta, S. Frintrop, and J. Hertzberg. Robust localization using context in omnidirectional imaging. In *Proceedings of IEEE International Conference on Robotics and Automation (ICRA)*, pages 2072–2077, Seoul, Korea, 2001.

[178] T. Palleja, M. Tresanchez, M. Teixido, and J. Palacin. Modeling floor-cleaning coverage performances of some domestic mobile robots in a reduced scenario. *Robotics and Autonomous Systems*, 58(1):37–45, December 2010.

[179] C. P. Papageorgiou, M. Oren, and T. Poggio. A general framework for object detection. In *Proceedings of 6th International Conference on Computer Vision (ICCV)*, pages 555–562, Bombay, India, 1998.

[180] N. Paragios and R. Deriche. Geodesic active regions and level set methods for supervised texture segmentation. *International Journal of Computer Vision*, 46(3):223–247, 2002.

[181] L. E. Parker. ALLIANCE: An architecture for fault tolerant multi-robot cooperation. *IEEE Transactions on Robotics and Automation*, 14(2):220–240, April 1998.

[182] L. E. Parker. Lifelong adaptation in heterogeneous multi-robot teams: Response to continual variation in individual robot performance. *Autonomous Robots*, 8(3):239–267, 2000.

[183] L. E. Parker. *Handbook of Robotics*, chapter Multiple Mobile Robot Systems, pages 921–941. Springer-Verlag, 2008.

[184] L. E. Parker, B. Kannan, X. Fu, and Y. Tang. Heterogeneous mobile sensor net deployment using robot herding and line-of-sight formations. In *Proceedings of IEEE/RSJ International Conference on Intelligent Robots and Systems (IROS)*, pages 2488–2493, Las Vegas, USA 2003.

[185] L. E. Parker, B. Kannan, F. Tang, and M. Bailey. Tightly-coupled navigation assistance in heterogeneous multi-robot teams. In *Proceedings of IEEE/RSJ International Conference on Intelligent Robots and Systems (IROS)*, volume 1, pages 1016–1022, Sendai, Japan, September 2004.

[186] D. Payton, R. Estkowski, and M. Howard. Pheromone robotics and the logic of virtual pheromones. In *Swarm Robotics. SAB 2004 International Workshop*, pages 45–57, Santa Monica, USA, 2004.

[187] D. P. Prasser, G. F. Wyeth, and M. J. Milford. Biologically inspired visual landmark processing for simultaneous localization and mapping. In *Proceedings of IEEE/RSJ International Conference on Intelligent Robots and Systems (IROS)*, pages 730–735, Sendai, Japan, 2004.

[188] F. P. Preparata and M. I. Shamos. *Computational Geometry: An Introduction*. Springer-Verlag, New York, USA, 1985.

[189] A. Pretto, E. Menegatti, Y. Jitsukawa, R. Ueda, and T. Arai. Image similarity based on discrete wavelet transform for robots with low-computational resources. *Robotics and Autonomous Systems*, 58(7):879–888, 2010.

[190] A. Ramisa, A. Tapus, R. L. de Mántaras, and R. Toledo. Mobile robot localization using panoramic vision and combinations of local feature region detectors. In *Proceedings of IEEE International Conference on Robotics and Automation (ICRA)*, pages 538–543, Pasadena, USA, 2008.

[191] I. Rekleitis, A. P. New, E. Rankin, and H. Choset. Efficient boustrophedon multi-robot coverage: an algorithmic approach. *Annals of Mathematics and Artificial Intelligence*, 52(2–4):109–142, April 2009.

[192] I. M. Rekleitis, G. Dudek, and E. Milios. Multi-robot collaboration for robust exploration. *Annals of Mathematics and Artificial Intelligence*, 31(1–4):7–40, 2001.

[193] J. Rittscher, J. Kato, S. Joga, and A. Blake. A probabilistic background model for tracking. In *Proceedings of European Conference on Computer Vision (ECCV)*, pages 336–350, Dublin, Ireland, 2000.

[194] E. Rosten, R. Porter, and T. Drummond. Faster and better: A machine learning approach to corner detection. *IEEE Transactions on Pattern Analysis and Machine Intelligence*, 32(1):105–119, 2010.

[195] J. A. Rothermich, M. I. Ecemiş, and P. Gaudiano. Distributed localization and mapping with a robotic swarm. In *Swarm Robotics. SAB 2004 International Workshop*, pages 58–69, Santa Monica, USA, 2004.

[196] A. Rowe, C. Rosenberg, and I. Nourbakhsh. A low cost embedded color vision system. In *Proceedings of IEEE/RSJ International Conference on Intelligent Robots and Systems (IROS)*, pages 208–213, Lausanne, Switzerland, 2002.

[197] H. A. Rowley, S. Baluja, and T. Kanade. Neural network-based face detection. *IEEE Transactions on Pattern Analysis and Machine Intelligence*, 20(1):23–38, 1998.

[198] Y. Rubner, C. Tomasi, and L. J. Guibas. The earth mover's distance as a metric for image retrieval. *International Journal of Computer Vision*, 40(2):99–121, 2000.

[199] S. Rutishauser, N. Correll, and A. Martinoli. Collaborative coverage using a swarm of networked miniature robots. *Robotics and Autonomous Systems*, 57(5):517–525, 2009.

[200] P. E. Rybski, S. T. Roumeliotis, M. Gini, and N. Papanikolopoulos. Appearance-based minimalistic metric SLAM. In *Proceedings of IEEE/RSJ International Conference on Intelligent Robots and Systems (IROS)*, pages 194–199, Las Vegas, USA, 2003.

[201] H. Sagan. *Space-Filling Curves*. Springer-Verlag, 1994.

[202] D. Scaramuzza and R. Siegwart. *Vision Systems: Applications*, chapter A Practical Toolbox for Calibrating Omnidirectional Cameras, pages 297–310. Vienna, Austria, 2007.

[203] M. Schael. Texture defect detection using invariant textural features. In *Pattern Recognition, Proceedings of 23rd DAGM Symposium*, pages 17–24, Munich, Germany, 2001.

[204] M. Schael. Invariant texture classification using group averaging with relational kernel functions. In *Proceedings of 2nd International Workshop on Texture Analysis and Synthesis*, pages 129–133, Copenhagen, Denmark, 2002.

[205] B. Schiele and J. L. Crowley. A comparison of position estimation techniques using occupancy grids. In *Proceedings of IEEE International Conference on Robotics and Automation (ICRA)*, pages 1628–1634, San Diego, USA, 1994.

[206] H. Schulz-Mirbach. Invariant features for gray scale images. In *Mustererkennung, Proceedings of 17th DAGM Symposium*, pages 1–14, Bielefeld, Germany, 1995.

[207] J. T. Schwartz and M. Sharir. A survey of motion planning and related geometric algorithms. *Artificial Intelligence*, 37(1–3):157–169, December 1988.

[208] S. Se, D. Lowe, and J. Little. Mobile robot localization and mapping with uncertainty using scale-invariant visual landmarks. *International Journal of Robotics Research*, 21:735–758, 2002.

[209] S. Se, D. Lowe, and J. Little. Vision-based global localization and mapping for mobile robots. *IEEE Transactions on Robotics*, 21(3):364–375, 2005.

[210] A. Senior, A. Hampapur, Y.-L. Tian, L. Brown, S. Pankanti, and R. Bolle. Appearance models for occlusion handling. *Image and Vision Computing*, 24(11):1233–1243, 2006.

[211] D. Serby, S. Koller-Meier, and L. V. Gool. Probabilistic object tracking using multiple features. In *Proceedings of the 17th International Conference of Pattern Recognition (ICPR)*, pages 184–187, Cambridge, UK, August 2004.

[212] J. Shi and J. Malik. Normalized cuts and image segmentation. *IEEE Transactions on Pattern Analysis and Machine Intelligence*, 22(8):888–905, August 2000.

[213] S. Siggelkow. *Feature Histograms for Content-Based Image Retrieval*. PhD thesis, Institute for Computer Science, University of Freiburg, Germany, December 2002.

[214] S. Siggelkow and H. Burkhardt. Improvement of histogram-based image retrieval and classification. In *Proceedings of 16th International Conference on Pattern Recognition (ICPR)*, pages 367–370, Quebec, Canada, 2002.

[215] C. Silpa-Anan and R. Hartley. Localisation using an image-map. In *Proceedings of the Australasian Conference on Robotics and Automation*, Canberra, Australia, 2004.

[216] R. Sim and G. Dudek. Learning generative models of scene features. *International Journal of Computer Vision*, 60(1):45–61, 2004.

[217] R. Simmons, D. Apfelbaum, W. Burgard, M. Fox, D. An Moors, S. Thrun, and H. Younes. Coordination for multi-robot exploration and mapping. In *Proceedings of the National Conference on Artificial Intelligence (AAAI)*, Austin, USA, 2000.

[218] R. Simmons, S. Singh, D. Hershberger, J. Ramos, and T. Smith. First results in the coordination of heterogeneous robots for large-scale assembly. In *Proceedings of the International Symposium on Experimental Robotics (ISER)*, pages 323–332, Honolulu, USA, December 2000.

[219] R. C. Smith and P. Cheeseman. On the representation and estimation of spatial uncertainty. *The International Journal of Robotics Research*, 5(4):56–68, 1986.

[220] D. Spears, W. Kerr, and W. Spears. Physics-based robot swarms for coverage problems. *International Journal on Intelligent Control and Systems*, 11(3):124–140, 2006.

[221] S. V. Spires and S. Y. Goldsmith. Exhaustive geographic search with mobile robots along space-filling curves. In *Proceedings of the First International Workshop on Collective Robotics*, pages 1–12, Paris, France, 1998.

[222] C. Stauffer and W. E. L. Grimson. Adaptive background mixture models for real-time tracking. In *IEEE Conference on Computer Vision and Pattern Recognition (CVPR)*, pages 246–252, Fort Collins, USA, 1999.

[223] G. S. Sukhatme, J. F. Montgomery, and R. T. Vaughan. *Robot Teams: From Diversity to Polymorphism*, chapter Experiments with Aerial-Ground Robots, pages 345–368. AK Peters, 2001.

[224] M. J. Swain and D. H. Ballard. Indexing via color histograms. In *Proceedings of Third International Conference on Computer Vision*, pages 390–393, Osaka, Japan, 1990.

[225] H. Tamimi. *Vision-based Features for Mobile Robot Localization*. PhD thesis, Department of Computer Science, University of Tübingen, Germany, 2006.

[226] H. Tamimi, H. Andreasson, A. Treptow, T. Duckett, and A. Zell. Localization of mobile robots with omnidirectional vision using particle filter and iterative SIFT. *Robotics and Autonomous Systems*, 54(9):758–765, 2006.

[227] H. Tamimi, A. Halawani, H. Burkhardt, and A. Zell. Appearance-based localization of mobile robots using local integral invariants. In *Proceedings of 9th International Conference on Intelligent Autonomous Systems (IAS)*, pages 181–188, Tokyo, Japan, 2006.

[228] H. Tamimi, C. Weiss, and A. Zell. Appearance-based robot localization using wavelets-PCA. In *Autonome Mobile Systeme*, pages 36–42, Kaiserslautern, Germany, 2007.

[229] A. Tapus and R. Siegwart. Incremental robot mapping with fingerprints of places. In *Proceedings of IEEE/RSJ International Conference on Intelligent Robots and Systems (IROS)*, pages 2429–2434, Alberta, Canada, 2005.

[230] S. Thrun. A probabilistic online mapping algorithm for teams of mobile robots. *International Journal of Robotics Research*, 20(5):335–363, 2001.

[231] S. Thrun, W. Burgard, and D. Fox. *Probabilistic Robotics*. MIT Press, 2006.

[232] S. Thrun, D. Fox, W. Burgard, and F. Dellaert. Robust monte carlo localization for mobile robots. In *Artificial Intelligence*, volume 1-2, pages 99–141, 2000.

[233] S. Thrun and J. J. Leonard. *Handbook of Robotics*, chapter Simultaneous Localization and Mapping, pages 871–889. Springer-Verlag, 2008.

[234] S. Thrun and Y. Liu. Multi-robot SLAM with sparse extended information filers. In *Proceedings of the 11th International Symposium of Robotics Research (ISRR)*, pages 254–266, Sienna, Italy, 2003.

[235] A. Torralba, R. Fergus, and W. T. Freeman. 80 million tiny images: a large dataset for non-parametric object and scene recognition. In *IEEE Transactions on Pattern Analysis and Machine Intelligence*, volume 30, pages 1958–1970, November 2008.

[236] R. Y. Tsai. A versatile camera calibration technique for high-accuracy 3D machine vision metrology using off-the-shelf TV cameras and lenses. *IEEE Journal of Robotics and Automation*, 3(4):323–344, August 1987.

[237] I. Ulrich and I. Nourbakhsh. Appearance-based place recognition for topological localization. In *Proceedings of IEEE International Conference on Robotics and Automation (ICRA)*, pages 1023–1029, San Francisco, USA, April 2000.

[238] C. Valgren and A. Lilienthal. SIFT, SURF and seasons: Long-term outdoor localization using local features. In *Proceedings of the 3rd European Conference on Mobile Robots (ECMR)*, pages 253–258, Freiburg, Germany, 2007.

[239] C. Veenman, M. Reinders, and E. Backer. Resolving motion correspondence for densely moving points. *IEEE Transactions on Pattern Analysis and Machine Intelligence*, 23(1):54–72, 2001.

[240] P. Viola, M. Jones, and D. Snow. Detecting pedestrians using patterns of motion and appearance. *International Journal of Computer Vision*, 63(2):153–161, 2005.

[241] P. Vorst and A. Zell. A comparison of similarity measures for localization with passive RFID fingerprints. In *ISR/ROBOTIK 2010 (Proceedings of the joint conference of ISR 2010 (41st International Symposium on Robotics) and ROBOTIK 2010 (6th German Conference on Robotics))*, pages 354–361, Munich, Germany, 2010.

[242] I. A. Wagner, Y. Altshuler, V. Yanovski, and M. Bruckstein. Cooperative cleaners: A study in ant robotics. *The International Journal of Robotics Research*, 27(1):127–151, 2008.

[243] I. A. Wagner, M. A. Lindenbaum, and M. Bruckstein. Distributed covering by ant-robots using evaporating traces. *IEEE Transactions on Robotics and Automation*, 15(5):918–933, October 1999.

[244] I. A. Wagner, M. A. Lindenbaum, and M. Bruckstein. MAC vs. PC: Determinism and randomness as complementary approaches to robotic exploration of continuous unknown domains. *International Journal of Robotics Research*, 19(1):12–31, 2000.

[245] X.-Y. Wang, J.-F. Wu, and H.-Y. Yang. Robust image retrieval based on color histogram of local feature regions. *Multimedia Tools and Applications*, 49(2):323–345, 2010.

[246] C. Weiss. *Self-Localization and Terrain Classification for Mobile Outdoor Robots*. PhD thesis, Department of Computer Science, University of Tübingen, Germany, 2008.

[247] C. Weiss, A. Masselli, H. Tamimi, and A. Zell. Fast outdoor robot localization using integral invariants. In *Proceedings of the 5th International Conference on Computer Vision Systems (ICVS)*, Bielefeld, Germany, March 2007.

[248] C. Weiss, H. Tamimi, A. Masselli, and A. Zell. A hybrid approach for vision-based outdoor robot localization using global and local image features. In *Proceedings of IEEE/RSJ International Conference on Intelligent Robots and Systems (IROS)*, pages 1047–1052, San Diego, USA, October 2007.

[249] K. Williams and J. Burdick. Multi-robot boundary coverage with plan revision. In *Proceedings of IEEE International Conference on Robotics and Automation (ICRA)*, pages 1716–1723, Orlando, USA, 2006.

[250] S. B. Williams, G. Dissanayake, and H. Durrant-Whyte. An efficient approach to the simultaneous localisation and mapping problem. In *Proceedings of IEEE International Conference on Robotics and Automation (ICRA)*, pages 406–411, Washington, USA, 2002.

[251] J. Wolf, W. Burgard, and H. Burkhardt. Robust vision-based localization by combining an image retrieval system with monte carlo localization. *IEEE Transactions on Robotics*, 21(2):208–216, April 2005.

[252] C. Wren, A. Azarbayejani, T. Darrell, and A. Pentland. Pfinder: Real-time tracking of the human body. *IEEE Transactions on Pattern Analysis and Machine Intelligence*, 19(7):780–785, 1997.

[253] Z. Wu and R. Leahy. An optimal graph theoretic approach to data clustering: Theory and its application to image segmentation. *IEEE Transactions on Pattern Analysis and Machine Intelligence*, 15(11):1101–1113, 1993.

[254] B. Yamauchi. Frontier-based exploration using multiple robots. In *Proceedings of the Second International Conference on Autonomous Agents*, pages 47–53, Minneapolis, USA, 1998.

[255] A. Yilmaz, O. Javed, and M. Shah. Object tracking: A survey. *ACM Journal of Computing Surveys*, 38(4):1–45, 2006.

[256] A. Yilmaz, X. Li, and M. Shah. Contour based object tracking with occlusion handling in video acquired using mobile cameras. *IEEE Transactions on Pattern Analysis and Machine Intelligence*, 26(11):1531–1536, 2004.

[257] A. Zelinsky, R. A. Jarvis, J. C. Byrne, and S. Yuta. Planning paths of complete coverage of an unstructured environment by a mobile robot. In *Proceedings of International Conference on Advanced Robotics (ICAR)*, pages 533–538, Tokyo, Japan, 1993.

[258] W. Zhang and J. Kosečká. Image based localization in urban environments. In *Proceedings of Third International Symposium on 3D Data Processing, Visualization and Transmission (3DPVT)*, pages 33–40, Chapel Hill, USA, 2006.

[259] Z. Zhang. A flexible new technique for camera calibration. *IEEE Transactions on Pattern Analysis and Machine Intelligence*, 22(11):1330–1334, 2000.

[260] X. Zheng, S. Jain, S. Koenig, and D. Kempe. Multi-robot forest coverage. In *Proceedings of IEEE/RSJ International Conference on Intelligent Robots and Systems (IROS)*, pages 3852–3857, Alberta, Canada, 2005.

[261] C. Zhou, Y. Wei, and T. Tan. Mobile robot self-localization based on global visual appearance features. In *Proceedings of IEEE International Conference on Robotics and Automation (ICRA)*, pages 1271–1276, Taipei, Taiwan, September 2003.

[262] S. C. Zhu and A. Yuille. Region competition: Unifying snakes, region growing, and Bayes/MDL for multi-band image segmentation. *IEEE Transactions on Pattern Analysis and Machine Intelligence*, 18(9):884–900, 1996.